METHODS of THERMODYNAMICS

Howard Reiss

Department of Chemistry and Biochemistry
University of California, Los Angeles

DOVER PUBLICATIONS, INC.
Mineola, New York

Bibliographical Note

This Dover edition, first published in 1996, is an unabridged and slightly corrected republication of the work first published in 1965 by Blaisdell Publishing Company (a division of Ginn and Company), New York, as "A Blaisdell Book in the Pure and Applied Sciences."

Library of Congress Cataloging-in-Publication Data

Reiss, Howard.
Methods of thermodynamics / Howard Reiss.
p. cm.
Originally published: New York : Blaisdell Pub., [1965].
Includes bibliographical references and index.
ISBN 0-486-69445-3 (pbk.)
1. Thermodynamics. I. Title.
QC311.R34 1996
536′.7—dc20 96-35379
CIP

Manufactured in the United States of America
Dover Publications, Inc., 31 East 2nd Street, Mineola, N.Y. 11501

TO MY WIFE

Phyllis

Preface

THERMODYNAMICS IS A MATURE MEMBER of the group of modern scientific disciplines. It is therefore surprising to witness the occurrence in recent years of a proliferation of new texts on the subject. Why has the attention of so many writers been engaged? Is it because there has been an explosive growth of scientific activity and a concomitant demand for students well grounded in fundamentals? Although this may represent *one* of the underlying motives, I feel that more compelling reasons exist. Among these must be listed the almost certain truth that nobody (authors included) understands thermodynamics perfectly. The writing of a book therefore becomes a kind of catharsis in which the author exorcises his own demon of noncomprehension and prevents it from occupying the soul of another.

The difficulties of understanding are compounded by the fact that the boundaries of thermodynamics cannot be defined with perfect exactness. Ask a man to define Newtonian mechanics and he will usually render an answer which is consistent with the definitions advanced by a majority of his fellows. The same concurrence of viewpoint cannot be found in connection with thermodynamics. Part of the problem lies in the fact that its boundaries are not static in spite of its respectable age. Attempts which have been made to apply thermodynamics to nonequilibrium processes and to fluctuating systems represent just two of the assaults on its classical boundaries. Perhaps the greatest source of confusion has arisen in the relation of statistical mechanics to thermodynamics. In principle, the formulas of thermodynamics can be derived by statistical methods, and so we have witnessed a blending of the two subjects, which has strengthened the statistical approach at the expense of reducing thermodynamics to an almost menial role.

The boundaries of thermodynamics are further blurred by the fact that the experimenter must himself make judgments as to what constitutes an acceptable thermodynamic system. Thus, he must determine whether a given system is in equilibrium (without having at his disposal an unchallenged definition of equilibrium), and he must also determine what are the independent variables of state. The procedure for making such decisions usually consists of *assuming* that a given system may be treated by thermodynamic methods and that certain variables may be regarded as state variables. The experimental and theoretical consequences of these assumptions are then explored. If grave inconsistencies do not appear, and useful results are forthcoming, the methods of thermodynamics may be considered applicable. This method of proceeding inescapably possesses aspects of circular reasoning. It amounts to the assertion that thermodynamics works because it works! On the other hand, the retention of a degree of openmindedness is of considerable value in the avoidance of misunderstanding. If nothing else, it relieves the student of any compulsion to seek absolute comprehension in a domain where it does not exist.

A first glance at thermodynamics leads to a false impression concerning the ease with which it can be mastered. A superficial perusal reveals that the subject does not require great mathematical facility. Certain parts of it may be learned by rote, especially where repetitious techniques are employed. In contrast, the mastery of the related discipline of statistical mechanics requires greater mathematical proficiency, but the subject lends itself more easily to understanding.

The salient feature of thermodynamics is its independence from considerations of microscopic phenomena. It provides relations among macroscopic variables which are invariant to all changes in our thinking concerning atomic and molecular processes. This divorcement from atoms and molecules, however, deprives it of those models which are so useful for intuitive reasoning. With statistical mechanics the contrary is true. One may delight himself with physical models. Thus, it is not uncommon to think of molecules as billiard balls, or of liquids as possessing quasi-crystalline structures.

Statistical analyses provide more information on the behaviors of systems than do the corresponding thermodynamic ones. Because of this, and the opportunity to employ models, many have been led to use statistical methods when thermodynamics would suffice. This is lamentable because the absolute confidence which we are willing to commit to relationships arrived at by thermodynamic means exceeds that which we are willing to place in statistical reasoning. Statistical mechanics then becomes an imperfect crutch used to bolster one's understanding of a relatively perfect body of thermodynamics.

The vogue for blending the two subjects has produced the situation in which many courses in thermodynamics are no longer *thermodynamics* but *statistical thermodynamics*. Blending may not only reflect a desire to use the statistical method as a crutch, but a very real need for compacting the expanded body of scientific knowledge downward so that students can consume it early enough to learn what must be learned while they are in training.

The present author is not a purist. Furthermore, he has a profound respect for the value of statistical mechanics. On the other hand, a student should learn to use the method of thermodynamics in all its power, in application to any problem in respect to which it may prove useful. In view of the high level of confidence which we place in thermodynamics, what is known thermodynamically is often considered to be known once and for all. Within the limits of this point of view, by restricting oneself initially to purely thermodynamic arguments, one can know what he *does* know before entering domains where conclusions are less certain. It is a pity, therefore, to discover that students who can manipulate thermodynamics with acceptable proficiency in areas to which they have seen it applied previously are often at a loss to deal with it in areas which are new. For example, students who can work effectively with systems in which surface is of no account often become confused when surface appears as a new variable.

This book is not intended as a textbook for a first course in thermodynamics. It is appropriately titled *The Methods of Thermodynamics*, and focuses attention on the physical technique of the subject rather than upon the entire corporeal substance. Emphasis will be placed on the problem areas of understanding. It should prove useful as an auxiliary text to be used along with any one of several standard treatments.

Almost all books on thermodynamics contain some errors which are not purely typographical. In this book, every attempt has been made to minimize this possibility. It should be pointed out that the present work is not the first attempt to provide a discussion of this sort. Every book on the subject makes some effort in this direction. In 1960, A. B. Pippard made an effective effort in his book *Elements of Classical Thermodynamics* (published by the Cambridge University Press). Insofar as books devoted to the conceptual nature of thermodynamics are concerned, the present author feels that Pippard's is the best.

The present book is similar to Pippard's in the degree of attention committed to method and concept. It differs from Pippard's in that the viewpoint is more that of the chemist than the physicist. Furthermore, whereas Pippard sought, among other things, to discuss certain advanced topics in classical thermodynamics, the emphasis in this book is focused not so much on that which is advanced as it is devoted to the clear explanation of what is elementary.

The plan of the book involves a gradual ascent towards a goal which appears in the middle rather than at the end. This goal is *the clear exposition of the significance and use of the thermodynamic potential.* The understanding of this concept and all of its ramifications represents the primary stumbling block for the student who would apply thermodynamics to very general systems possessing all sorts of unconventional variables. The thermodynamic potential can only be understood when there exists a clear recognition of the role of *constraints* in thermodynamics. For this reason, much of this book is devoted to the discussion of constraints.

Clear understanding of the nature of the thermodynamic potential makes possible the specification of a general recipe for the construction of such potentials for the most general of systems. With this recipe in hand, the summit will have been passed; and the rest of the book is an easy descent, during which the technique of application of the recipe is demonstrated in connection with specific systems.

Exhaustive treatments of such special systems are avoided, and only enough examples are given to clothe the bones with enough flesh to make the organism viable. Our concern remains with the *method* rather than the detailed application of the subject.

As far as the detailed application is concerned, the reader may consult any one of a large number of texts. For a comprehensive treatment of the application of thermodynamics to chemical problems, the reader is directed to the revision by K. S. Pitzer and L. Brewer of the classic, *Thermodynamics* (McGraw-Hill Book Co., Inc., 1961), written originally by G. N. Lewis and M. Randall. For the detailed application of thermodynamics to problems of interest to engineers and physicists, the book *Heat and Thermodynamics* (McGraw-Hill, 1957) by M. W. Zemanski is recommended. For the treatment of special advanced topics, there is Pippard's book, mentioned above, and in addition a book by H. B. Callen entitled *Thermodynamics* (John Wiley & Sons, Inc., 1960).

Callen's book is interesting in that it is an example of the postulational approach applied in a rather pure form. It contains some excellent discussions of special topics—for example, the thermodynamics of elastic systems.

Several other good detailed treatments of thermodynamics are available. One of these is *Chemical Thermodynamics* by J. G. Kirkwood and I. Oppenheim (McGraw-Hill, 1961). Another, also called *Chemical Thermodynamics* is by I. Prigogine and R. Defay (John Wiley, 1962). Older books which have merit are: *Thermodynamics and Chemistry* by F. H. MacDougall (John Wiley, 1939); and *Textbook of Thermodynamics* by P. S. Epstein (John Wiley, 1937). Finally, there is *Thermodynamics* by E. A. Guggenheim (North-Holland Publishing Co., 1959).

With the exception of Pippard's book, all these texts are more or less comprehensive treatments, and yet they manage to cover different areas of the subject. This is an indication of the scope which an all-embracing treatise would have to encompass.

In keeping with my belief that thermodynamics should be mastered for its own sake, extensive reference to statistical mechanics will be avoided in the present book. Any mention of molecular theory will be purely incidental and should not contribute to the general line of argument.

It will be noticed that the treatment of ionic systems, and especially of ionic systems in electric fields, has been avoided, although this is a popular subject in most works on chemical thermodynamics. The author has followed a deliberate policy in omitting such systems from consideration because they cannot be treated (at least not as recognizably ionic systems) without relying heavily on molecular theory.

This book has had the benefit of constructive criticism from many of my colleagues. I am especially indebted to Dr. E. R. Cohen and Professor W. G. McMillan, both of whom read the manuscript thoroughly and offered detailed suggestions, many of which have been incorporated in the text. Others who should be mentioned are J. M. Zimmerman, S. J. Yosim, L. E. Topol, C. Warner, T. Wolfram, and still others too numerous to mention at this time. I am particularly indebted to Mrs. Rita Finger for her untiring assistance in the preparation of the manuscript; and to my wife, Phyllis, who, with stoicism, made the usual sacrifice of domestic order so that this book could be written. Last but not least, I am grateful to North American Aviation, Inc., for providing me with the time and environment necessary for the successful completion of the task.

HOWARD REISS

Thousand Oaks, California
April 1964

Contents

I. Some General Concepts

1. Objectives of Thermodynamics 1
2. The Thermodynamic System 2
3. Equilibrium 3
4. Thermodynamic State, Variables of State 4
5. Macroscopic State Space 5
6. Mechanical Work 6
7. Quasistatic Processes 9
8. Correspondence between Constraints, Variables, and Work 11
9. Metastable Equilibrium 16
10. The Form of Modern Science 18

II. Mathematical Apparatus

1. Exact Differentials and Pfaff Differential Expressions 20
2. Theorem of Carathéodory 22
3. Transformation of Variables 26
4. Decomposition of a Partial Derivative 27
5. Euler's Theorem and Homogeneous Functions 28
6. Constrained Extremals 29

III. The First Law of Thermodynamics

1. Laws in Thermodynamics 32
2. Temperature 33
3. Temperature Scales 36
4. Adiabatic Work 37

5. Internal Energy, the First Law, and Heat 41
6. Heat Capacity, Enthalpy, and Heat of Change 43
7. Phases 45
8. Intensive and Extensive Quantities 46
9. Euler's Theorem and Partial Molar Quantities 47

IV. The Second Law and Entropy

1. General Remarks 50
2. Need for an Additional Extensive Function of State 51
3. The Degree of Constraint 54
4. Entropy 56
5. Extremal Properties of the Entropy 59
6. Virtual Variations 60
7. Temperature Scale and Thermodynamic Efficiency 62
8. Maximum Efficiency 65
9. Additional Use of the Reversible Environment 68
10. Conversion of Heat into Work 69
11. The Principle of Carathéodory 71
12. Efficiency in Engineering Problems 77
13. The Helmholtz and Gibbs Free Energies 77
14. Legendre Transformations and Maxwell Relations 79

V. Ideal Substances

1. Equation of State 82
2. The Ideal Gas 83
3. Internal Energy of the Ideal Gas and Relation between the Kelvin and Thermodynamic Scales 85
4. Variation of the Entropy of an Ideal Gas 89
5. The Entropy of Mixing of Two Ideal Gases 90
6. Ideal Solutions 93
7. $\bar{G}_i$ for a Component of an Ideal Solution and Proof that the Volume of Mixing is Zero 95

VI. Some Useful Formulas

1. Compressibility and Expansivity and the Relation between C_v and C_p 99

2. Energetics of the Free Expansion of a Gas 101
3. The Joule-Thomson Coefficient 103
4. Relation of Enthalpy and Entropy to Heat Capacity 106
5. Magnetic Substances 106

VII. Internal Equilibrium and the Extremal Properties of the Entropy

1. Extremal Condition on the Entropy 108
2. A One-Component, Two-Phase System 109
3. "Feature" of Equilibrium 112
4. Internal Potentials 113

VIII. Thermodynamic Potentials

1. Concept of the Thermodynamic Potential 115
2. An Elementary Method 116
3. Sign of the Work Performed by the Variational Constraint 118
4. The Sign of $-\Delta w_{1_{rev}}$ 120
5. Thermodynamic Potentials and Extremal Conditions 122
6. Generalized Feature of Equilibrium 125
7. Alternative Representations of the Chemical Potential 127
8. Proof of the Conditions of Internal Equilibrium using either U, H, A, or G 129
9. Generalization to any Number of Phases with any Number of Components 131
10. The Phase Rule 132
11. Chemical Potential as a Partial Molar Quantity 133
12. Open Systems 134

IX. Phase Equilibria in Simple Systems

1. Dependence of Chemical Potential on Variables of State 136
2. The Clapeyron-Clausius Equation 137
3. Raoult's Law 139
4. Boiling Point Elevation and Freezing Point Depression 140
5. Chemical Equilibrium and the Law of Mass Action 142
6. Thermodynamic Activity, Dilute Solutions 143
7. The Measurement of G, Electrochemical Cells 146

X. Osmotic Systems

1. More on the Generalized Feature of Equilibrium 150
2. Osmotic Systems 151

XI. Systems Which May Perform Surface Work

1. Surface Layers 156
2. The Laplace Relation 157
3. The Thermodynamic Potential 160
4. An Alternative Form for $R^{(2)}$ 161
5. The Chemical Potential 163
6. Dependence of Vapor Pressure on Drop Size 165
7. Dependence of Solubility on Drop Size 167
8. Surface Effects and Structure of the Surface Layer 167
9. Excess Quantities in General 169
10. Surface Constraints and Surface Tension 170
11. The Thermodynamic Potential 172
12. The Chemical Potential 174
13. The Gibbs-Duhem Relation for the Surface 175
14. The Gibbs Adsorption Isotherm 177

XII. Systems in Gravitational and Centrifugal Fields

1. Application of the Generalized Feature of Equilibrium 178
2. Chemical Potential in an External Field 180
3. Role of the External Field 182
4. The Barometric Formula 184
5. Centrifugal Fields 185

XIII. Elastic Systems

1. Stress and Strain 187
2. Thermodynamic Potential of a Simple Isotropic System 188
3. The Chemical Potential 190
4. Vapor Pressure of an Elastic System 191

XIV. Stability

1. Stable Equilibrium 194
2. Mechanical Stability 195
3. Thermal Stability 199
4. Compositional Stability 200
5. Direction of Flow 201
6. Equilibrium between Drop and Vapor 202

XV. The Third Law

1. Introductory Remarks 204
2. Cooling 204
3. The Nernst Postulate 207
4. Specific Heats and Various Derivatives at Absolute Zero 208
5. The Modified Nernst Postulate 209
6. The Unattainability of Absolute Zero 210

INDEX 215

Methods
of
Thermodynamics

I

Some General Concepts

1. Objectives of Thermodynamics

Throughout this book, it will be assumed that the reader has had rudimentary thermodynamic training such as one might acquire in an elementary course in physical chemistry. In later chapters, we shall take time to define fundamental parameters such as temperature, surface tension, et cetera, whose aggregate forms the discipline of thermodynamics; but for the moment, we shall deal with the subject discursively rather than rigorously, assuming that the reader has some familiarity with these concepts. Later we shall deal with these same subjects more thoroughly.

Thermodynamics deals with the properties of matter in bulk. Although we know that matter is composed of elementary particles and may have a detailed organized structure, we cannot call upon this information for the purpose of thermodynamic reasoning. This does not mean that it is not possible nor desirable to use extra-thermodynamic information in conjunction with thermodynamic methods, but only that one should be careful to isolate thermodynamic from nonthermodynamic reasoning, especially because those relationships which have been established by means of thermodynamics will be invariant to all changes in our concepts of atomic and molecular phenomena. We shall refer to systems of matter in bulk as *macroscopic*, whereas atoms and molecules constitute *microscopic* systems.

At this point in time, one is usually willing to place a very high level of confidence in conclusions arrived at by thermodynamic means. Actually, there are not many things in this world which come to us bound with immutable fact; and therefore, we are led to suspect that there are not many things which can be known by thermodynamics alone. This is true. To make thermodynamics useful, it will be necessary to feed a great deal of information into its mechanisms. Our profit will be other facts which might have been obtained through the agency of additional measurement. This is one of the major services of thermodynamics—the avoidance of redundant measurements among macroscopic variables. For example, suppose that the coefficient of thermal expansion, the isothermal compressibility, the heat capacity at constant volume, and the temperature and volume of a substance have been measured. It will then prove unnecessary to measure its heat capacity at constant pressure, for thermodynamics reveals that the heat capacity is entirely determined by the former quantities. In another example, suppose that the latent heat of vaporization of a liquid and the heat capacities of both the liquid and its vapor have been measured. If the vapor pressure is known at one temperature, it becomes possible to specify its value at other temperatures through the application of thermodynamic methods.

This is the primary service afforded by thermodynamics, namely a means of transforming certain useful macroscopic data concerning a system into other useful macroscopic data on the same system. There are other applications. None of them, however, is quite so general as the one just mentioned. For example, thermodynamic methods have been utilized to investigate the stability of systems against change; or they have been used by engineers to discuss efficiencies, that is, the efficiency with which heat can be converted into mechanical work.

Originally, thermodynamics was intended to provide relationships between the parameters which describe systems at *equilibrium*. More recently, attempts have been made to apply thermodynamic methods to the macroscopic parameters of nonequilibrium systems. This endeavor has been rewarded with some, although limited, success. It has been suggested that what was originally thermodynamics should now be called *thermostatics*, and that the term thermodynamics should be reserved for the methods used to treat nonequilibrium systems. The general methods employed to interrelate macroscopic data in the manner described are conveniently referred to as *phenomenological*. Thus, thermostatics and thermodynamics are phenomenologies. In this book, the original meaning of the word thermodynamics will be retained, and the phenomenology of nonequilibrium processes may be denoted by the term *irreversible* thermodynamics.

2. The Thermodynamic System

We have already employed words like "experimental data," "system," "equilibrium," and "temperature" without adequate definitions. This is acceptable for purely discursive purposes such as these brief introductory passages, but will prove entirely inadequate in what follows. Although we shall employ the discursive style from time to time throughout the remainder of this chapter, there is something to be gained by defining a few things carefully. It is logical to begin with the definition of the thermodynamic system.

> A thermodynamic system is an arbitrary geometrical portion of the universe with fixed or movable boundaries which may contain matter or energy or both.

Thus, a thermodynamic system is merely that part of the universe which we elect (arbitrarily) to focus attention upon. All the rest of the universe is then considered to be the environment.

3. Equilibrium

We have mentioned that thermodynamics in its classical sense deals with systems which are in *equilibrium*. The definition of equilibrium requires careful discussion. Generally, there will be certain macroscopic attributes of the system (thermodynamic properties) which can be measured and assigned numerical values. Such properties might be pressure, temperature, volume, density, heat capacity, et cetera, most of which also require careful definition. Having understood the nature of thermodynamic properties we are in a position to define equilibrium:

> A thermodynamic system is in equilibrium when none of its thermodynamic properties are changing with time at a measurable rate.

This definition leaves a lot to be desired, but it is difficult to furnish a better one. It provides the experimenter with a certain latitude in his assessment of what constitutes equilibrium. Its utility rests upon the idea that a continuity of phenomena exists such that if the system is not truly at equilibrium, and changes are occurring at an infinitesimal rate, then the relationships between macroscopic parameters will differ only infinitesimally from the relationships at true equilibrium.

The definition, unless further qualified, leaves the door open to still further confusion. For example, *steady state* phenomena are known in which matter or energy are flowing through the system at a definite rate while the local thermodynamic properties remain unchanged. The steady conduction of heat through a solid in which the local distributions of temperature and density remain invariant is an example of such a process. Such *steady* states must be distinguished from true equilibrium even though they seem consistent with the definition presented above. Usually the experimenter will have no difficulty in recognizing them, since obvious changes will take place in the surroundings of the thermodynamic system, even though the appearance of the system itself may remain invariant.

Another source of confusion which may arise in connection with the above definition concerns the reproducibility of the equilibrium state. Once it has been ascertained that the properties of the system are not changing with time and that a steady state is not involved, it may still be impractical to treat the system by the methods of thermodynamics. Before thermodynamics can be applied effectively, the experimenter must be able to reproduce the state at will. To accomplish this, he must know what variables determine the state; and even more important, he must be able to control these variables. As an example, a solid may be prepared in many states which meet the requirement of invariance to time (at least during the time of a measurement). But each of these states may be complicated by the fact that the solid contains an array of internal strains. To each distinct set of strains will correspond (under our definition) another state of equilibrium. The strains themselves are variables of the system, and unless we can control them and reproduce them, it will not be practical to employ the methods of thermodynamics.

4. Thermodynamic State, Variables of State

We are now in a position to define a *thermodynamic state*, a term which, unless otherwise specified, will be restricted to the description of systems at equilibrium. If the thermodynamic properties of a system at equilibrium are observed, it will be noted that whenever a certain number of them assumes certain numerical values all the remaining properties take on perfectly definite values. For example, in the case of a substance like water, whenever a given mass† has a certain temperature and pressure, the volume of that mass will be determined in the sense that whenever the values of temperature and pressure are reproduced so will be the value of its volume. Alternatively, whenever the temperature and volume assume certain values, the pressure will be perfectly well defined. In order for all the properties to be reproduced, it is only necessary that the same *number* of them be reproduced. This number is called the *degrees of freedom*, or the number of *independent variables* of the system. A thermodynamic state is then defined in terms of the independent variables, each set of values for the independent variables corresponding to a state. For a macroscopic system, the number of independent thermodynamic variables is usually small. By contrast, if it were necessary to describe the microscopic state of some thermodynamic system, say a mole of water, it would be necessary to specify the values of the coordinates of position and momentum for all the elementary particles constituting that mole. Thus, the number of independent microscopic variables might be as high as 10^{24} or 10^{25}.

† Sometimes a thermodynamic state is defined intensively (see Section 8, Chapter 3) so that, for example, water at a given temperature and pressure is considered to be in the same state independent of its mass. In this book, "open" systems (see Section 5, Chapter 3 and Section 12, Chapter 8) will be considered, and it is important to distinguish, as different states, conditions of different mass.

Why is so profound a simplification achieved in proceeding from the microscopic to the macroscopic mode of description? The answer (at least a crude one) seems to lie in the existence of reproducible average behavior. The rate of microscopic motion is so high that during the time required for the performance of a single macroscopic thermodynamic measurement the system has passed through a near infinity of microscopic states. At best, therefore, the thermodynamic measurement can sense the time average behavior of the system as it passes through its multitude of microscopic states. The *average* behavior, however, is reproducible and is determined by the specification of only a few conditions. These are the thermodynamic variables of state.

Specification of the number of independent variables, like the specification of equilibrium, must be accomplished by the experimenter. Thus, a given system having fixed composition in bulk may have temperature and volume as independent variables. Suppose now that it is pulverized into a very fine powder, or, if it is a liquid, dispersed into fine droplets. The reproducibility of the state of the system may then depend upon the experimenter's ability to reproduce its surface area as well as its temperature and volume. Under this condition there are three rather than two independent variables. At some state of subdivision the experimenter would have to decide to take account of the surface. This will be necessary whenever he discovers that reproducing the temperature and volume of the system does not guarantee the reproduction of a given state. When the experimenter decides that a property must be treated as an independent variable, he must in effect *assume* that this is the case, and his assumption should be a direct outgrowth of experiment.

This example may be used to point up the fact that the experimenter will have to use his judgment in deciding when a variable or property is macroscopic. Thus, if a process of subdividing is continued, the relevant surface area can become the total molecular surface (insofar as it is possible to give meaning to this concept). The total molecular surface will not do as a thermodynamic variable. Only macroscopic properties will suffice, and the experimenter must decide when a property ceases to be macroscopic.

5. Macroscopic State Space

Thermodynamics, taking no account of atomic or molecular structure, does not lend itself to discussion in terms of models and may therefore appear to be overly vague. Since models can provide valuable assistance to analytical thinking, it is worthwhile to introduce what is probably the next best thing in the abstraction of thermodynamics, namely the *macroscopic state space*. This is an abstract hyperspace whose coordinate axes are the several independent thermodynamic variables of the system in question. Thus, the state space of a simple one-component substance of fixed mass might be a plane along whose two cartesian axes are plotted the temperature and volume of the system. Any thermodynamic state can then be represented by a point on this plane, the coordinates of which are the temperature and volume appropriate to the state.

This example involves a two-dimensional state space. If, in addition, the mass of the substance is permitted to vary, then mass becomes a third independent variable; and it is necessary to augment the state space with a third cartesian axis. Then we have a three-dimensional state space. If there are more than three independent variables, more axes must be added so that a hyperspace of more than three dimensions results.

State space provides a framework within which, in lieu of any models, one can perform thermodynamic reasoning in a more concrete manner. Notice that all the points in state space, since they represent thermodynamic states, are equilibrium points. Any two points in this space specify a thermodynamic *change* of state. Notice that the *change* of state does not specify the path. There are many different paths which correspond to the same change of state; that is, the same end points may be connected by a plurality of paths. In fact, it may not be possible to plot some paths in state space. These are nonquasistatic (see Section 7) paths which do not consist of sequences of equilibrium states. We will employ the term *process* to describe a path by which two points in state space are connected. Thus, to a given *change of state* there may correspond several processes, whereas to a given *process* there corresponds only one change of state.

The reader will find it a great aid to his understanding if, whenever possible, he performs his thermodynamic reasoning in terms of state space. In our discussions throughout the book, we shall return to it again and again until it becomes a realistic domain within which we are made to feel at home.

6. Mechanical Work

The expeditious application of thermodynamics to various systems requires us to be able to express, quantitatively, the *mechanical work* exchanged between a system and its environment. In fact, it will emerge as the text unfolds that the method of treatment of a given system depends, more than anything else, upon the *kind* of work which the system may perform.

If the system is a simple substance, liquid water for example, then it may perform work on its environment, or the environment may perform work on it, when its volume changes. Suppose that the medium surrounding the system is a fluid which exerts only hydrostatic stresses at the pressure p. Then if the volume of the system is V and an infinitesimal change in volume is represented by dV, the work performed *by* the system on its environment may be represented by†

$$Dw = pdV. \tag{1.1}$$

This relation is easily derived if the liquid is considered to be enclosed by a piston in a cylinder of cross-sectional area A, and if its volume is increased by dV through the outward movement of the piston through the distance dl. The force on the piston is simply pA, the pressure times the area, and the work done (force times distance) is therefore

$$Dw = pAdl = pdV, \tag{1.2}$$

since $dV = Adl$. Equation (1.1) may be generalized to a system of any shape.[1]

Most of the time no difficulty is encountered in arriving at a quantitative expression such as (1.1). For example, in the case of an elastic system the incremental work performed on the elastic body may be represented as the force required to induce the extension multiplied by the incremental extension; or in the case of a system in which surface is important, the work performed in extending the surface may be expressed as the surface tension multiplied by the incremental extension in surface. These forms are all quite similar to Equation (1.1).

In a slightly different example, consider the electrical work performed by an electrochemical cell (which may also be treated as a thermodynamic system). If the potential difference across the terminals of the cell is denoted by $\mathscr{E}$, then when dQ units of charge are driven by the cell from one terminal to another the work performed is

$$Dw = \mathscr{E}\, dQ. \tag{1.3}$$

Sometimes, however, it is not easy to arrive at a quantitative expression for the work. Consider, for example, a substance in a magnetic field. Great care must be exercised in specifying the work of magnetization. It may depend upon the configuration of the magnetic field, the shape of the substance, and whether or not the material is diamagnetic, paramagnetic, or ferromagnetic. Detailed discussions of this problem may be found in books by A. B. Pippard,[2] H. B. Callen,[3] and K. S. Pitzer and L. Brewer.[4] For

† The use of the large D in Dw instead of the customary symbol for the differential is explained in Section 5, Chapter 3.

[1] J. G. Kirkwood and I. Oppenheim, *Chemical Thermodynamics* (New York: McGraw-Hill Book Co., Inc., 1961), pp. 14–15.
[2] A. B. Pippard, *Elements of Classical Thermodynamics* (New York: Cambridge University Press, 1960), pp. 24–26.
[3] H. B. Callen, *Thermodynamics* (New York: John Wiley & Sons, Inc., 1960), pp. 238–243.
[4] K. S. Pitzer and L. Brewer, *Thermodynamics* (revision of the original text by G. N. Lewis and M. Randall) (New York: McGraw-Hill Book Co., Inc., 1961), pp. 497–512.

the special case of a nonferromagnetic, isotropic specimen placed in a uniform magnetic field (such as might be found within a long solenoid) the incremental work performed on the system may be expressed by[2–4]

$$-Dw = d\left[\frac{\mathscr{H}^2 V}{8\pi}\right] + \mathscr{H}\, d\mathscr{M}. \tag{1.4}$$

(Here we have used $-Dw$ instead of Dw for the work symbol because Equation (1.4) represents work performed *on* the system rather than work performed *by* the system as in Equation (1.3).) The quantity $\mathscr{H}$ represents the *external* magnetic field (unrationalized units have been used), that is, the magnetic field present in the solenoid *before* the introduction of the ellipsoid, and $\mathscr{M}$ is the magnetization. It should be emphasized that Equation (1.4) is applicable only to the special system under consideration. For details, the reader should consult page 512 of Reference 4. Excellent insight into the matter may be acquired by consulting Reference 2.

The first term on the left of Equation (1.4) represents the work which must be performed in increasing the external field throughout the volume V occupied by the specimen. The second term more properly represents the work of polarizing the specimen. The total field is actually composed of a superposition of the external field and that generated by the polarized body. In this sense, some of the energy stored in the *total* field is stored outside of the geometrical confines of the magnetic specimen. As we have indicated in Section 2, a thermodynamic system consists of a geometric region, and the work which the system does is measured operationally in terms of the mechanical energy exchanged across the defineable boundaries of this region. This limitation causes some difficulty in the direct measurement of the work of magnetic polarization.

It is customary to arbitrarily subtract the first term in Equation (1.4), leaving only

$$-Dw = \mathscr{H}\, d\mathscr{M} \tag{1.5}$$

as the incremental work performed in increasing the magnetic moment by $d\mathscr{M}$. However, this separation is not strictly thermodynamic since no method exists for measuring $\mathscr{H}\, d\mathscr{M}$ without at the same time measuring the first term of Equation (1.4). Thus, one cannot distinguish by macroscopic experimental means the modes of energy storage within the volume V of the magnetic specimen.

In the end the distinction must be made by theoretical methods involving continuum electromagnetic theory. Nevertheless, in Chapters 6 and 14 we shall follow this procedure (since it is conventional), taking care to remember our reservations.

Problems of similar nature are encountered in dealing with electrically polarizable systems, and the reader is referred to Reference 2, page 27; Reference 3, pages 243–245; and Reference 4, pages 447–512 for further details.

7. Quasistatic Processes

Only equilibrium points may be plotted in state space. Thus, in order to be representable in state space, any path must pass through a sequence of equilibrium states. It is obvious that any real change between thermodynamic states must take place at a finite rate so that the intermediate states cannot be equilibrium ones. Such a path cannot be represented in state space. However, if the change is made to occur more and more slowly, then in a limiting way the path converges on some path in state space. Over the limiting path the change is said to take place in a *quasistatic* or *reversible* manner.

The first appellation is more transparent than the second, and we shall have more to say below about the precise implication of the word "reversible." For the moment, it is sufficient to regard it as descriptive of any process which can be plotted in state space. Even though a truly reversible process cannot in fact occur (by definition equilibrium precludes change), the concept is useful. It is still possible to calculate certain attributes of a reversible process even though it cannot occur. At the very least, its path can be plotted in state space. But more often it is possible to do other things, such as computing the amount of work that would be performed by a system during a reversible change. Generally it is possible to do this when the *equation of state* of the system is known. We shall have more to say about the equation of state at a later time; but for the moment, we take it to mean the relationship between the pressure, volume, and temperature of a simple one-component system.

The equation of state for one mole of an *ideal* gas (see Chapter 5) is

$$pV = RT, \tag{1.6}$$

where p is the pressure, V the volume, T the temperature, and R the gas constant. These quantities will be discussed more carefully later. If the independent variables are chosen to be V and T, then a thermodynamic change of state can be defined, for example, by the two states having values for these variables (V_1,T_1) and (V_2,T_1), respectively. This change is isothermal; no change in temperature is involved. The work performed along a path in state space connecting these points is (according to Equation (1.1))

$$\Delta w = \int_{V_1}^{V_2} p\, dV; \tag{1.7}$$

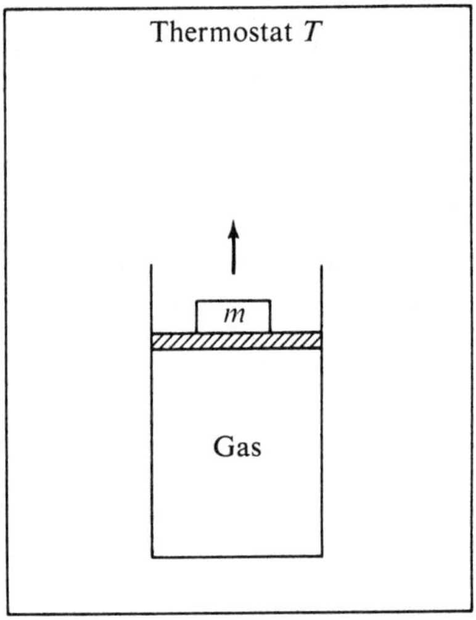

FIGURE 1.1. *Gas enclosed by weighted piston in thermostated cylinder.*

it being understood that the temperature is fixed at T_1. Substitution of Equation (1.6) into Equation (1.7) then yields

$$\Delta w = RT_1 \ln \frac{V_2}{V_1}. \tag{1.8}$$

Since the path is in state space, this is work performed during a reversible change. The nomenclature *reversible* (which may not be so desirable as the term *quasistatic*) has its origin in the fact that at any stage an infinitesimal change of external conditions can cause the process to be reversed in every detail.

Consider the system in Figure 1.1. This consists of a gas contained in a vessel closed by a piston (cross hatched). On the piston rests a mass m which is just insufficient in magnitude to balance the pressure of the gas. Thus, the gas expands and the piston moves in the direction indicated by the arrow. The entire system is rendered isothermal by being immersed in a thermostat. Since the mass is just insufficient by an infinitesimal amount to balance the pressure of the gas, the latter expands very slowly, in fact quasistatically. As a result, the piston does not dissipate energy by frictional processes for these vanish at vanishing velocities. Neither does the gas dissipate energy by internal friction. The reason is the same.

If the expansion were to be conducted rapidly, the state of the gas would be very nonuniform. Any hydrostatic pressure would be nonuniform, and there would be even more general stresses which would vary from point to point in the gas. Similar difficulties would be experienced in connection with the temperature. If m failed to balance the pressure of the gas by an appreciable amount, a rapid expansion would result. Similarly, a rapid compression would result if at any stage an increment in m were made large enough to exceed the gas pressure by a large amount. In this case, the only feature of clear reversal in the process would be the change from expansion to compression. The various frictional processes mentioned above would not necessarily reverse, nor would the processes leading to the nonuniformities in temperature and pressure be exactly reversed.

On the other hand, in the quasistatic process in which m exceeds the gas pressure by an infinitesimal amount, the pressure and temperature are always defined (the process is a sequence of equilibrium states and can be plotted in state space). Furthermore, there is a gradual flow of heat from the thermostat to the gas to maintain the temperature as the gas does work against the mass m. If the mass is now increased by an infinitesimal amount, the quasistatic expansion will give way to a quasistatic compression. Every process will be completely reversed. For example, besides the change from expansion to compression the flow of heat from the thermostat to the gas will be reversed. It is this reversal in every aspect of the process which gives rise to the term *reversible*.

We re-emphasize the fact that a reversible process can be defined as one which can be plotted in state space.

8. Correspondence between Constraints, Variables, and Work

Every state of equilibrium is subject to certain constraints which are imposed upon the system. In fact, constraints and variables of state are in one-to-one correspondence. This point can be made clear through reference to several examples. Consider the simple system of Figure 1.2. This consists of a rigid container immersed in a thermostat so that the entire system is maintained at a fixed temperature. The container is divided into two compartments by a movable partition which can be locked in place. Each compartment contains a gas of the same substance. The partition is locked in place so that the pressures of the two gases are different, with p_2—the pressure of gas 2—being less than p_1—the pressure of gas 1. The state of this system is now described by three independent variables. These could be chosen as the common temperature T and the two pressures p_1 and p_2. Alternatively, they might be chosen as T, one pressure p_1, and a *number* specifying the position of the locked partition (or the volume of one compartment).

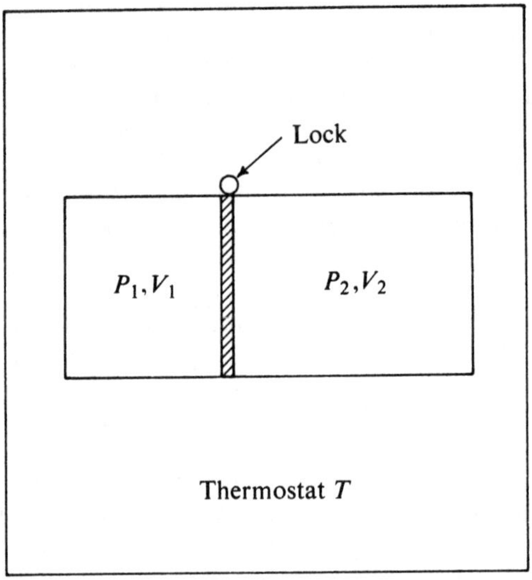

FIGURE 1.2. *Illustrating the removal of a constraint; in this case the release of a partition separating two gases at different pressures.*

The various constraints on the system are the thermostat maintaining the fixed temperature, the fixed volume of the container, the lock on the partition, and so forth. If the partition is unlocked so that it can move freely, we know that it will move until the pressures are equalized, so that $p_2 = p_1$. Under this condition, the independent variables of the system will only be two in number; for example, T and one pressure, say p_1. The removal of the constraint has therefore resulted in the loss of an independent variable.

This situation is quite general. Constraints are always associated with variables. The addition of a constraint implies the addition of an independent variable. Other examples are easily thought of. Consider a reactive chemical system in solution in a beaker. When this system comes to equilibrium the independent variables will usually be temperature, pressure, and certain variables of composition. Suppose now that the same system is arranged as an electrochemical cell and that an external electromotive force is impressed across this cell. Then a certain amount of electrolysis can be made to occur. If the cell is of the right kind, the electrolysis will generate a new state of equilibrium in which the composition of the system is different from that which corresponded to equilibrium in the beaker. Now, however, the independent variables will be temperature, pressure, variables of composition, and the value of the impressed electromotive force. The additional constraint of the impressed electromotive force has added a new variable.

Consider another example. Suppose that we have two solutions separated by a semipermeable membrane, the entire system being enclosed in a thermostat so that the temperature remains the same throughout. For definiteness, assume that solution 1 is of pure water and that solution 2 consists of sucrose dissolved in water at concentration c_2. For the sake of symmetry, assign a concentration c_1 to solution 1. Of course, since this solution consists of pure water, $c_1 = 0$. Suppose that the semipermeable membrane passes water but not sucrose. Then it is well known that osmosis will occur, and that equilibrium will only be established when the pressure p_2 on solution 2 exceeds the pressure p_1 on solution 1 by a prescribed amount.

The variables of state which now describe this equilibrium are four in number. They may be chosen in a variety of ways. For example, we can choose the set c_1, c_2, p_1, and T, the common temperature. Alternatively, one may choose c_1, p_2, p_1, and T or c_2, p_2, p_1, and T. The semipermeability of the membrane represents a constraint forbidding the passage of sucrose. The relaxation of this constraint permits the establishment of an equilibrium in which $c_1 = c_2$ and $p_1 = p_2$. Under these conditions, the independent variables are only three in number. They can be chosen as the set c_1, p_1, and T. Once more, the removal of a constraint has led to the loss of a variable.

Consider still another example. Suppose a vessel contains a mixture of three gases: hydrogen, oxygen, and water vapor. If there are appreciable amounts of hydrogen and oxygen present, it is well known that it is possible for a considerable amount of chemical reaction to occur. However, the activation energy for this reaction is so high that under ordinary circumstances, at reasonable temperatures, the reaction proceeds very slowly. It can be made to occur violently by the addition of a catalyst. A constraint is effectively present, preventing the occurrence of reaction.

To describe fully the thermodynamic state of this system when the reaction cannot occur, it is necessary to specify the temperature, volume, and the partial pressures of each of the three constituent gases. Thus, there are five independent variables of state in all. When the constraint is lifted, as for example by the introduction of a platinum catalyst, the reaction proceeds to some point of equilibrium. This new equilibrium, however, is determined by four variables. For now it is only necessary to specify the partial pressures of two of the gases as well as the temperature and volume of the mixture. The relation implicit in the attainment of chemical equilibrium has eliminated a variable. Thus, once again the removal of a constraint leads to the removal of a variable.

Just as constraints and variables are in one-to-one correspondence with one another, they are each in one-to-one correspondence with the *kinds* of work which a system may perform on its environment or vice versa. Thus, consider the example of Figure 1.2. When the lock is released, the partition moves under the pressure differential $p_1 - p_2$. If the partition is coupled (like an automobile piston) to an engine, it may perform useful work. In this case, the work is clearly volume work. The energy is delivered directly to the partition (in the form of motional energy) so that one might consider the work to be performed against the agency of constraint. This point of view may be summarized by the statement that work is performed *against* the constraint. If after the pressures have been equalized one wishes to reapply the original constraint, the partition will have to be driven to the left (against the pressure differential) so that work will be performed *on* the system. In this case, one might say that in the achievement of the *constraint* work is performed on the system, or more briefly (but less accurately), that the constraint performs work on the system.

The point to be made is that the *kind* of work (in this case *volume* work at a given pressure) depends upon the *arrangement* of the system which is in turn related to the constraints which are applied. Consider the example, discussed above, of the chemical reaction set up alternatively, in a beaker, or more elaborately as an electrochemical cell. When set up, in a simple manner, in a beaker, the volume change which accompanies the occurrence of the chemical transformation results in the performance of *volume* work, either by the system on its environment, or vice versa, depending upon whether an increase or decrease of volume is involved. On the other hand, when the system is arranged as a chemical cell and an external emf is applied (as a constraint), the system may exchange *electrical* work with its environment. Work may be performed on the environment when the external emf is the back emf of a motor which the cell is driving; or alternatively, electrical work may be performed on the cell (causing some electrolysis) if the motor is reversed and employed as a dynamo. Thus, the additional constraint, the external emf, makes possible the performance of an additional kind of work, namely electrical work.

It should be borne in mind that, in the cell, the chemical transformation which attends the performance of electrical work is still accompanied by a change in volume which gives rise to volume work. Thus, the original system may perform two kinds of work simultaneously, provided that two appropriate constraints are involved. The situation may be generalized to any number of constraints with a corresponding number of kinds of work. A system may be continually augmented so that new constraints, new variables, and new kinds of work are possible.

In this connection, it should be pointed out that a system comes to equilibrium *subject* to certain constraints. For convenience, we will denote the set of constraints, subject to which the system attains equilibrium, by the symbol χ_0. As an example, consider once more the chemical reaction discussed above. If the chemical solution is enclosed in a thermostat at temperature T and the environmental pressure is maintained at p while the system is prevented from exchanging matter with its surroundings, the reaction will proceed until a definite state of equilibrium is reached. This equilibrium will be determined by T, p, and the relative amounts of the various components brought together to form the solution in the first place. These conditions form the set of constraints χ_0, subject to which the final state of equilibrium is achieved.

Now if none of these constraints are removed, and (assume the system has been set up as a cell) an external emf of magnitude $\mathscr{E}$ is applied, further changes can be made to occur. For example, the reaction may be forced to travel over the path inverse to the one which it followed in arriving at equilibrium subject to the set of constraints χ_0, that is, some electrolysis may occur. If the chemical system does not undergo a phase transition (the solution does not become saturated with a given component and precipitation is avoided), it is a fact of experience that equilibrium will once more be established, but now with proportions of reactants and products different from the properties characteristic of the first equilibrium subject to χ_0. In fact, the new proportions, and therefore the new state of equilibrium, will be determined not only by the set of constraints χ_0 but by the new constraint implicit in the emf of magnitude $\mathscr{E}$. As we have indicated above, this new equilibrium will be characterized by a new variable as well as a new constraint.

Thus, it may be seen in the special case of the electrochemical cell that even though an initial state of equilibrium may be attained subject to an initial set of constraints χ_0, it is possible to establish a new state of equilibrium without relaxing any of the constraints χ_0 by merely introducing a new constraint, or new constraints (in this case the emf), which we may designate by the symbol χ_1. The new equilibrium is then established subject to the augmented set of constraints $\chi_0 + \chi_1$.

Although we have used the electrochemical cell as a convenient example, the situation is quite general; and any thermodynamic system which has achieved equilibrium subject to an initial set of constraints χ_0 may be displaced to a new state of equilibrium with the aid of a new constraint (or set of constraints) χ_1. In the new state, however, the system will possess more independent variables.

The situation is somewhat similar to the familiar one found in mechanics in connection with a pendulum which is subject initially only to the force of gravity, and therefore assumes a state of equilibrium in which it hangs vertically downward. This state is illustrated in Figure 1.3, in which the initial state is labeled state 0. The force of the pendulum arm reacting to gravity may be regarded as the constraint χ_0 which determines this initial state. A new state of equilibrium may be produced through the introduction of a displacing force (shown in Figure 1.3) which drives the system to state 1 on the right. The displacing force is the analogue of the additional constraint χ_1, and the new state is determined by both χ_0 and χ_1.

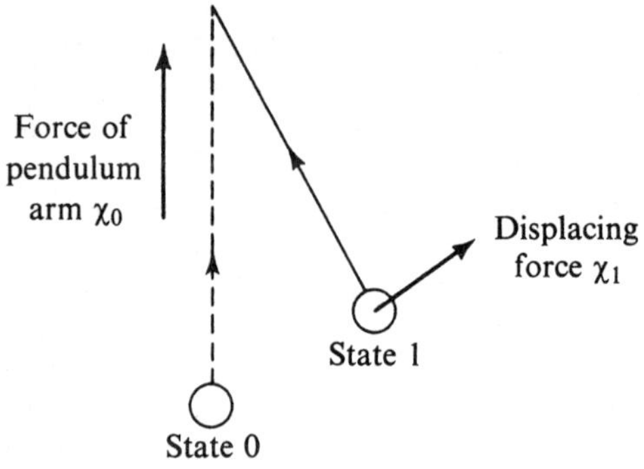

FIGURE 1.3. *Illustrating the displacement of a pendulum through the application of a force χ_1. The initial state under the "constraint" of the pendulum arm χ_0 is state* 0. *The displacement drives the system to a new equilibrium state, state* 1, *which is determined by both the constraints χ_0 and χ_1.*

It will be noticed that χ_1 must perform work on the pendulum if the displacement is to be achieved. Also, work may be performed by the pendulum arm if its length is made variable (for example, if it were somewhat elastic). Thus, work may be exchanged with both sets of constraints, the initial set χ_0 and the displacing set χ_1, during the displacement. The same is true in respect to a thermodynamic system—work is exchanged with both χ_0 and χ_1. The work performed against χ_0 may be denoted by Δw_0 and that performed against χ_1 may be denoted by Δw_1.† The total work performed *on* the system during the displacement is therefore $-(\Delta w_0 + \Delta w_1)$.

Since the thermodynamic system moves from a state controlled by, say, n variables to one controlled by $n + 1$ or more variables, the change would be represented in macroscopic state space (if it were made to occur reversibly) by a path which begins on an n-dimensional hypersurface and moves off this surface into an $n + 1$ (or higher) dimensional space.

9. Metastable Equilibrium

There is an interesting point associated with the osmotic example and the example of the hydrogen, oxygen, and water mixture of the last section which can lead to confusion unless properly clarified. This has to do with the concept of *metastable* equilibrium. Conventionally, the mixture of

† We are here using the symbol Δ to denote a finite amount of work Δw in contrast to our earlier use of D to denote an infinitesimal quantity of work Dw.

hydrogen, oxygen, and water vapor in the absence of platinum is said to be in *metastable* equilibrium. This is meant to imply that the equilibrium is in some way incomplete.

Now our definition of thermodynamic equilibrium leaves no room for the distinction between metastable and stable equilibrium, since it only demands that the macroscopic properties of the system do not change at a measurable rate. An imperceptible rate of change on the molecular scale, such as the slow chemical reaction in the case of our gas mixture, does nothing to alter this. The system must still be counted as being at equilibrium. The introduction of the word "metastable" therefore adds a complication, even an inconsistency, which can be confusing and which one might well do without. In the strict sense, all equilibria are stable (or unstable—see footnote on page 18) provided the system is assigned the correct number of variables. The only distinction between the uncatalyzed metastable equilibrium and the catalyzed stable equilibrium has to do with the number of variables. The catalyzed system is stable when referred to four variables, whereas the uncatalyzed is metastable when referred to this number but stable when referred to five.

The concept of metastability seems to have been employed mainly to indicate the degree of control the experimenter has over the constraint.† For example, he has great control over the lock of Figure 1.1 or over the electromotive force impressed across the cell described in Section 8 of this chapter, but very little control over the reaction in the last example. He cannot easily turn it only partly off. Nevertheless, these degrees of control

† An example in which the experimenter's degree of control is even less than that which may be exercised over the hydrogen-oxygen-water mixture concerns a system such as supercooled water. In the case of hydrogen mixed with oxygen, the experimenter may prepare the system with many different compositions (in fact the composition may be varied continuously by the method of preparation) and the reaction will not occur. As a result, the system may actually be carried over a path of chemical change which can be plotted in state space, that is, a quasistatic path. It is even possible to measure *volume* reversible work associated with this change by measuring the intermediate changes in the volume of the system. The *total* reversible work corresponding to this change cannot, however, be *directly* measured since some of it is performed against the constraint that prevents the spontaneous reaction along the indicated path, and this constraint is not clearly defined. In the case of supercooled water, the reactant is water and the resultant is ice. The variable of composition (which corresponds to the constraint which stabilizes supercooled water) is the ratio of the mass of ice to that of water. In this case, arbitrary intermediate mixtures of ice and water in contact cannot be prepared because the presence of the ice catalyzes the freezing of the supercooled water.

However, from the thermodynamic point of view, differences in degree of control are meaningless. One could imagine a hypothetical constraint which would stabilize a mixture of ice and supercooled water in any arbitrary proportions. One can even achieve intermediate compositions by keeping the ice and water in separate containers (provided that interfacial effects between ice and water are of negligible importance). As with the hydrogen-oxygen case, it is possible to measure the *volume* work associated with the *reversible* transformation of supercooled water to ice. One need only measure the volumes of the intermediate mixtures (ice and water being confined to separate containers) and determine the volume work from the incremental volumes.

are indistinguishable for the purposes of thermodynamic reasoning. The fact that the reaction does proceed imperceptibly in the absence of a catalyst in no way invalidates this point of view.

This is emphasized by the osmotic system in the last example. Here a pressure differential is established across the membrane. The stability of the membrane against this differential is in fact one of the constraints on the system. Now membranes under such pressure differentials can flow, but so slowly that the process is imperceptible. This process plays the same role in the osmotic system as the chemical reaction involving hydrogen and oxygen plays in the chemical system. We have no difficulty in viewing osmotic phenomena as manifestations of stable equilibria, yet the situation is logically identical to the case of hydrogen and oxygen in the absence of a catalyst.

Thus, in the interest of clarity it is preferable to describe all equilibrium states which can be achieved by a system as stable,† the proper number of variables always being specified. Every equilibrium is therefore subject to constraints, and it is impossible to describe an equilibrium without at the same time prescribing the constraints to which it is subject.

10. The Form of Modern Science

Modern scientific disciplines are being cast increasingly into deductive frameworks. This, in a sense, is an admission that it is the major task of science to describe rather than to explain. Euclidean geometry, which describes spatial relationships on a plane, provides a good example of such a deductive framework. There would be nothing to prevent one from wandering about the plane world with protractor and ruler performing all sorts of measurements. If the results of all these measurements were recorded, an enormous thesis would result. This compendium would be nothing more than a detailed quantitative description of plane spatial relations. It could hardly be called an explanation.

Now it is actually possible to present the same description in a more efficient manner. There is a central essence in the immense collection of facts recorded by the above-mentioned measurements. Instead of recording the measurements themselves, it is more convenient to record this central essence. Thus, we arrive at the laws of plane geometry. These laws are themselves just facts, unprovable by logical syllogism, but once assumed true they can serve as the basis for the proof, by logic alone, of the truth of all the other facts.

† Another class of equilibrium which is not *stable*, but also not *metastable*, exists. This class, which will be discussed in Chapter 14, is called *unstable* equilibrium. It is related to the concept of *stable* equilibrium, and we shall not elaborate it at this point.

In Euclidean geometry, an example of a typical law or "axiom" is the assertion "Things equal to the same thing are equal to each other." This law, although unprovable, seems eminently reasonable. From it, and a few other laws, we can derive the theorems of geometry. Each theorem purports to describe the result of a possible measurement, and collectively the theorems describe spatial relationships on the plane. Obviously the laws stand or fall depending upon how well the theorems actually describe the results of measurement. It is important to note that nothing further in the way of explanation has been achieved by the adoption of the deductive approach. The efficiency of description has, however, been increased, since it is all contained in the few positive statements which constitute the set of laws. Familiar sets of disciplinary laws are the laws of mechanics, the laws of electricity and magnetism, the laws of quantum mechanics, and especially the laws of thermodynamics.

Thermodynamics, like geometry, may be presented in deductive form. A very formal axiomatic approach may be adopted in which the laws are presented in highly mathematical language. A good example of this method of presentation may be found in H. B. Callen's book (Reference 3). Although there are certain advantages to this approach (and no loss of generality), it tends not to expose with maximum efficiency the physical basis of the subject; and the present author prefers to travel a middle course in which the laws of thermodynamics are presented in axiomatic form, but where detailed discussions are also presented in order to illuminate the physical basis of the assumptions.

One very great advantage of the formal axiomatic approach (in which the laws are presented in highly mathematical language) is that it enables one to see logical similarities among the laws. One can *see how* the laws make the thermodynamic method possible and *how* each additional law renders it more powerful.

§ II

Mathematical Apparatus†

1. Exact Differentials and Pfaff Differential Expressions

Although no very elegant mathematical apparatus is required for the expeditious use of thermodynamics, some mathematical tools are necessary, and we shall discuss these in this chapter. No attempt at rigor will be made, and in most cases we shall be content with little more than the exposition of certain relationships without proof. The student who is interested in further details may consult any text on advanced calculus.

We begin our discussion with the consideration of exact differentials and Pfaff expressions. Consider a real function $F(x,y,z,\ldots)$ of the variables $x, y, z, \ldots$ which is analytic except perhaps at certain isolated singular points. The set $x, y, z, \ldots$ may be the state variables of a thermodynamic system, in which case the axes along which they are plotted constitute the macroscopic state space of the system. $F(x,y,z,\ldots)$ is then a *function of state*. The total differential of F, namely,

$$dF = \left[\frac{\partial F}{\partial x}\right]_{y,z,\ldots} dx + \left[\frac{\partial F}{\partial y}\right]_{x,z,\ldots} dy + \left[\frac{\partial F}{\partial z}\right]_{x,y,\ldots} dz + \ldots, \quad (2.1)$$

is called an *exact* or *complete* differential because it is the differential of some unique analytic function of $x, y, z, \ldots$. (Since F is analytic, it possesses continuous higher order derivatives except at the isolated singular points.) If dF is integrated over some path (whose path length element is defined by dx, dy, dz, et cetera) connecting two points 1 and 2 in state space, the result is simply $F(x_2,y_2,z_2,\ldots) - F(x_1,y_1,z_1,\ldots)$. This quantity depends only on the points 1 and 2 and not on the particular path connecting them. This is

† This chapter is included for the convenience of those readers with scanty mathematical background. For those whose mathematical training is more extensive, it is possible to omit the chapter entirely. Those who wish to use it may refer to it as particular mathematical tools become necessary to support the thermodynamic development. For this purpose appropriate references are made as the text unfolds and therefore it is possible to skip to Chapter 3 without loss of continuity.

an important characteristic of an exact differential. Its path integral is a function only of the end points of the path and not of the path itself.

Not every differential expression of the form

$$D\psi = X\,dx + Y\,dy + Z\,dz + \ldots, \tag{2.2}$$

where X, Y, Z, . . . are analytic functions of x, y, z, . . . , is the differential of a single valued analytic function of x, y, z, . . . (an exact differential). The symbol $D\psi$ is designed to make it clear that $D\psi$ may not be the differential of some function ψ. This is why ψ is preceded by a capital D. $D\psi$ is nothing more than a shorthand notation for the expression on the right. We reserve the lower case d for complete differentials.

Since by definition

$$dF = X\,dx + Y\,dy + Z\,dz + \cdots \tag{2.3}$$

is a complete differential, comparison of Equation (2.3) with Equation (2.1) reveals that

$$X = \left[\frac{\partial F}{\partial x}\right]_{y,z,\ldots}, \quad Y = \left[\frac{\partial F}{\partial y}\right]_{x,z,\ldots}, \quad Z = \left[\frac{\partial F}{\partial z}\right]_{x,y,\ldots}. \tag{2.4}$$

It may be shown, as a consequence of F being analytic, that

$$\left[\frac{\partial}{\partial y}\left[\frac{\partial F}{\partial x}\right]_{y,z,\ldots}\right]_{x,z,\ldots} = \left[\frac{\partial}{\partial x}\left[\frac{\partial F}{\partial y}\right]_{x,z,\ldots}\right]_{y,z,\ldots}, \tag{2.5}$$

or from Equation (2.4)

$$\left[\frac{\partial X}{\partial y}\right]_{x,z,\ldots} = \left[\frac{\partial Y}{\partial x}\right]_{y,z,\ldots}. \tag{2.6}$$

Therefore, it makes no difference whether F is differentiated with respect to x or y first. Similar *reciprocity* relations exist for other pairs of independent variables besides (x,y).

Expressions of the kind (2.2) are usually called *Pfaff* differential expressions. It can be shown that the necessary and sufficient condition for $D\psi$ to be an exact differential dF is simply that a relation of the kind (2.6) holds for every pair of independent variables. Frequently, when $D\psi$ is not an exact differential, it can be made so through division by a function $\lambda(x,y,z,\ldots)$ called an *integrating denominator*. In this case it is possible to write

$$dF = \frac{D\psi}{\lambda}. \tag{2.7}$$

2. Theorem of Carathéodory

When does a Pfaff differential *expression* possess an integrating denominator? To answer this we introduce the Pfaff differential *equation* $D\psi = 0$. Thus, the Pfaff differential equation corresponding to the expression (2.2) is (for three independent variables)

$$D\psi = X\,dx + Y\,dy + Z\,dz = 0. \tag{2.8}$$

Equations usually determine something, and it is natural to inquire into what a Pfaff equation determines.

Suppose X, Y, and Z are interpreted as the x, y and z components of a vector $\mathbf{R}$,

$$\mathbf{R} = \mathbf{i}X + \mathbf{j}Y + \mathbf{k}Z, \tag{2.9}$$

where $\mathbf{i}$, $\mathbf{j}$, and $\mathbf{k}$ are unit vectors along the x, y, and z axes. The differentials dx, dy, and dz can also be interpreted as the components of a vector

$$\mathbf{d}r = \mathbf{i}\,dx + \mathbf{j}\,dy + \mathbf{k}\,dz. \tag{2.10}$$

Then Equation (2.8) may be written in the form

$$\mathbf{R} \cdot \mathbf{dr} = 0, \tag{2.11}$$

where the dot stands for the usual scalar product.

Equation (2.11) implies that the vectors $\mathbf{R}$ and $\mathbf{dr}$ are perpendicular to one another. Now $\mathbf{R}(x,y,z)$ is well defined at each point in the space of $\mathbf{R}$ because X, Y, and Z are well defined. On the other hand, $\mathbf{dr}$ is determined by Equation (2.11) only to the extent that it lies in a plane perpendicular to $\mathbf{R}$ at each point (x,y,z). The vector $\mathbf{dr}$ can be thought of as defining the direction of a curve lying, at each point in $\mathbf{R}$ space, in a direction normal to the local direction of the vector $\mathbf{R}$. Obviously there is a plurality of curves which are determined in this manner by Equation (2.8).

Thus, it is clear that Pfaff differential equations specify certain curves in space. These may be called *solution curves* associated with the differential equation (2.8).

Suppose $D\psi$ *is* an exact differential dF. Then comparison of Equations (2.4) and (2.9) reveals that

$$\mathbf{R} = \text{grad } F. \tag{2.12}$$

Now the gradient of F is always perpendicular to some member of the uniparametric family of nonintersecting surfaces,

$$F(x,y,z) = K, \tag{2.13}$$

where K is the parametric constant. Therefore considerations of continuity require all curves which are perpendicular to grad F at a point to lie in one such surface. In fact, the uniparametric set of surfaces generated by varying K contains all solution curves. Thus, when $D\psi$ is an exact differential, the solution curves lie in the surfaces generated by setting the function, whose differential is involved, equal to a parametric constant.

When $D\psi$ is not an exact differential, the set of solution curves satisfying Equation (2.8) cannot be used to generate families of surfaces. The simplest example of this is the Pfaff equation for the so-called *null complex*:

$$D\psi = y\,dx - x\,dy - \frac{\omega}{2\pi}\,dz = 0, \tag{2.14}$$

where ω is a constant. Comparison with Equation (2.2) reveals that the components of the vector **R** are now

$$X = y, \qquad Y = -x, \qquad Z = -\frac{\omega}{2\pi}. \tag{2.15}$$

Consider the following equations:

$$z = \frac{\omega}{2\pi}\arcsin\frac{y}{a} \tag{2.16}$$

$$x^2 + y^2 = a^2. \tag{2.17}$$

These describe a parametric family of spirals of radius (and parameter) a with axis along the z-axis, and with pitch ω. By differentiation of Equations (2.16) and (2.17) we find

$$\frac{dy}{dx} = -\frac{x}{y} \tag{2.18}$$

$$\frac{dz}{dx} = -\frac{\omega}{2\pi y}. \tag{2.19}$$

Thus, the vector pointing along the direction of the spiral has components X, Y, and Z which stand in the following ratios to each other:

$$\frac{Y}{X} = \frac{dy}{dx} = -\frac{x}{y} \tag{2.20}$$

$$\frac{Z}{X} = \frac{dz}{dx} = -\frac{\omega}{2\pi y} \tag{2.21}$$

or

$$X:Y:Z = y : -x : -\frac{\omega}{2\pi}. \tag{2.22}$$

Comparison of Equation (2.22) with Equation (2.15) indicates that the vector **R** corresponding to the *null complex* points along the direction of one of the spirals defined by Equations (2.16) and (2.17).

Now for $a = 0$, the spiral degenerates to the z-axis and since **R** then points along the z-axis the solution curves of Equation (2.14) run perpendicular to that axis. On the other hand, even for an infinitesimal nonzero value of a, they must run perpendicular to the direction of the curve of a spiral and therefore decidedly nonperpendicular to the z-axis. It should be evident that such a group of solution curves cannot be used to define a single surface.

Correspondingly, $D\psi$ in Equation (2.14) is not an exact differential. Thus, applying the test of Equation (2.6),

$$\frac{\partial X}{\partial y} = \frac{\partial y}{\partial y} = 1 \neq \frac{\partial Y}{\partial x} = \frac{\partial(-x)}{\partial x} = -1. \tag{2.23}$$

If the situation is specialized to the two-variable case then

$$D\psi = X\,dx + Y\,dy = 0 \tag{2.24}$$

or

$$\frac{dy}{dx} = -\frac{X}{Y}.$$

This can be viewed as an ordinary differential equation for $y(x)$; and since X and Y are well behaved a solution always exists, except perhaps at those singular points where $Y = 0$. This solution can be represented as a uniparametric family of nonintersecting curves

$$y = y(x,K) \tag{2.25}$$

or

$$F(x,y) = K,$$

where K is the "integration constant" obtained when solving Equation (2.24). This is the two-dimensional analogue of the uniparametric set of surfaces (2.13).

The total differential of F obtained from Equation (2.25) is

$$dF = \left[\frac{\partial F}{\partial x}\right]_y dx + \left[\frac{\partial F}{\partial y}\right]_x dy = 0. \tag{2.26}$$

The family of solution curves defined by Equation (2.25) or equivalently (2.26) were obtained from the Pfaffian form (2.24). Hence, comparing Equation (2.26) with Equation (2.24), it must follow that dF may differ from $D\psi$ only by a multiplicative factor, say $\frac{1}{\lambda}$. Thus,

$$dF = \frac{D\psi}{\lambda}, \tag{2.27}$$

where λ is an integrating denominator. Note that λ will usually be a function $\lambda(x,y)$ of x and y. We thus arrive at the important but *special* result that

> a Pfaff differential expression in two variables always possesses an integrating denominator.

The same result does not hold for expressions in three or more variables. On the other hand, we are in a position to draw an important conclusion involving a necessary *consequence* of the existence of an integrating denominator no matter how many variables are involved. We notice that in the case of two variables where an integrating denominator is assured, the solution curves are nonintersecting because of the uniqueness of solutions of Equation (2.24). Thus:

> In an arbitrary neighborhood of any point P, there are points P' which cannot be connected to P by a solution curve.

This is clear because the nonintersecting solution curves form a continuum. Thus, if a point P and an arbitrary neighborhood of P are given, it is always possible to find a point P' in the neighborhood, lying on a different solution curve from the one containing P. Since P and P' are on different solution curves, they cannot be connected by a solution curve, for this would violate the nonintersection property. Thus, in the two-variable case, if a Pfaff differential *expression* has an integrating denominator then:

> Arbitrarily close to a point P there are other points P' which cannot be reached starting at P over a path which is a solution curve of the corresponding Pfaff differential equation.

This can be generalized to the case of three or more variables. We have already demonstrated in the three-variable case that the existence of an integrating denominator for a Pfaff expression requires that the Pfaff equation have solution curves which lie within nonintersecting surfaces (see Equation (2.13)). Thus, the above argument for nonintersecting curves can be generalized to the case of two nonintersecting surfaces. We thus arrive at the general conclusion:

> If a Pfaff differential expression in any number of variables is known to have an integrating denominator, then in the domain of these variables, arbitrarily close to any point P, there are other points P' which are inaccessible from P over a path which is a solution curve of the corresponding Pfaff differential equation.

It is natural to ask whether the converse is true; that is, if inaccessible points exist in the neighborhood of every point, does the Pfaff expression possess an integrating denominator? The answer is in the affirmative, and an elegant proof has been given by Carathéodory.[5,6] Although the proof is not difficult, we shall give only a statement of the result here.

If a Pfaff differential expression

$$D\psi = X\,dx + Y\,dy + Z\,dz + \cdots$$

has the property that in the space of its variables every arbitrary neighborhood of a point P contains other points which are inaccessible from P along a path corresponding to the solution of the equation $D\psi = 0$, then an integrating denominator for the expression exists.

3. Transformation of Variables

Suppose F to be a function of only three variables, x, y, z. (The following discussion applies to functions of any number of variables, but it is convenient to illustrate the point with only three.) Suppose further that x, y, z can be expressed as functions of three other variables, u, v, w;

$$\begin{aligned} x &= x(u,v,w), \\ y &= y(u,v,w), \\ z &= z(u,v,w). \end{aligned} \tag{2.28}$$

Then through use of these equations it is possible by direct substitution to express F as a function of u, v, w; that is,

$$F\{x(u,v,w),\ y(u,v,w),\ z(u,v,w)\} \equiv G(u,v,w).$$

In thermodynamic manipulation it is customary to sacrifice the mathematically-precise mode of expression of the function independent of whether x, y, z or u, v, w are regarded as the independent variables. Thus, instead of using $G(u,v,w)$ we use $F(u,v,w)$, the change in functional form being noted but not expressed. Now depending upon whether F is expressed as $F(x,y,z)$ or $F(u,v,w)$ we may write

$$\begin{aligned} dF &= \left[\frac{\partial F}{\partial x}\right]_{y,z} dx + \left[\frac{\partial F}{\partial y}\right]_{x,z} dy + \left[\frac{\partial F}{\partial z}\right]_{x,y} dz \\ &= \left[\frac{\partial F}{\partial u}\right]_{v,w} du + \left[\frac{\partial F}{\partial v}\right]_{u,w} dv + \left[\frac{\partial F}{\partial w}\right]_{u,v} dw. \end{aligned} \tag{2.29}$$

[5] C. Carathéodory, *Math. Am.*, 67 (1909), 355.
[6] M. Born, *Physik. Z.*, 22 (1921), 249.

The various partial derivatives are interrelated by the "chain rule" of elementary differential calculus. Thus,

$$\left[\frac{\partial F}{\partial u}\right]_{v,w} = \left[\frac{\partial F}{\partial x}\right]_{y,z}\left[\frac{\partial x}{\partial u}\right]_{v,w} + \left[\frac{\partial F}{\partial y}\right]_{x,z}\left[\frac{\partial y}{\partial u}\right]_{v,w} + \left[\frac{\partial F}{\partial z}\right]_{x,y}\left[\frac{\partial z}{\partial u}\right]_{v,w},$$

$$\left[\frac{\partial F}{\partial v}\right]_{u,w} = \left[\frac{\partial F}{\partial x}\right]_{y,z}\left[\frac{\partial x}{\partial v}\right]_{u,w} + \left[\frac{\partial F}{\partial y}\right]_{x,z}\left[\frac{\partial y}{\partial v}\right]_{u,w} + \left[\frac{\partial F}{\partial z}\right]_{x,y}\left[\frac{\partial z}{\partial v}\right]_{u,w}, \quad (2.30)$$

$$\left[\frac{\partial F}{\partial w}\right]_{u,v} = \left[\frac{\partial F}{\partial x}\right]_{y,z}\left[\frac{\partial x}{\partial w}\right]_{u,v} + \left[\frac{\partial F}{\partial y}\right]_{x,z}\left[\frac{\partial y}{\partial w}\right]_{u,v} + \left[\frac{\partial F}{\partial z}\right]_{x,y}\left[\frac{\partial z}{\partial w}\right]_{u,v},$$

with similar inverse relations for $\left[\frac{\partial F}{\partial x}\right]_{y,z}$, et cetera, in terms of derivatives with respect to u, v, and w. These equations of transformation, as well as Equation (2.6), are of great importance in the manipulations of thermodynamics.

By returning to Equation (2.29) we may introduce Equation (2.4) in order to obtain

$$dF = X\,dx + Y\,dy + Z\,dz. \quad (2.31)$$

In these terms, the first of Equations (2.30) becomes

$$\left[\frac{\partial F}{\partial u}\right]_{v,w} = X\left[\frac{\partial x}{\partial u}\right]_{v,w} + Y\left[\frac{\partial y}{\partial u}\right]_{v,w} + Z\left[\frac{\partial z}{\partial u}\right]_{v,w}. \quad (2.32)$$

Thus, as a formal rule (an aid to memory) we may regard Equation (2.32) as derived from Equation (2.31) through division by the differential du of the new independent variable—only the differentials dF, dx, dy, and dz being affected by the new denominator. The same mnemonic may be employed, of course, in dealing with the second and third of Equations (2.30).

4. Decomposition of a Partial Derivative

Another useful relation can be derived from Equation (2.1) specialized to the case of two variables,

$$\partial F = \left[\frac{\partial F}{\partial x}\right]_y dx + \left[\frac{\partial F}{\partial y}\right]_x dy. \quad (2.33)$$

Consider only those values of x and y which satisfy the relation

$$F(x,y) = \text{constant}. \quad (2.34)$$

This relation defines y as a function $y(x)$ of x so that $dy = \left[\frac{dy}{dx}\right]_F dx$. Hence, using Equation (2.34) we find

$$dF = \left[\frac{\partial F}{\partial x}\right]_y dx + \left[\frac{\partial F}{\partial y}\right]_x dy = \left[\frac{\partial F}{\partial x}\right]_y dx + \left[\frac{\partial F}{\partial y}\right]_x\left[\frac{\partial y}{\partial x}\right]_F dx = 0, \quad (2.35)$$

and therefore it must follow that

$$\left[\frac{\partial F}{\partial x}\right]_y = -\left[\frac{\partial F}{\partial y}\right]_x \left[\frac{\partial y}{\partial x}\right]_F, \tag{2.36}$$

a relation which can often be used to advantage in thermodynamic manipulations.

5. Euler's Theorem and Homogeneous Functions

An important theorem which plays a wide role in thermodynamics is Euler's theorem on homogeneous functions. A homogeneous function is defined as follows. Consider $F(x,y,z,\ldots)$. If α and n are constants and

$$F(\alpha x,\alpha y,\alpha z,\ldots) = \alpha^n F(x,y,z,\ldots), \tag{2.37}$$

then F is said to be a homogeneous function of degree n in the variables $(x,y,z,\ldots)$.

If we define

$$\alpha x = x',\ \alpha y = y',\ \alpha z = z',\ \text{et cetera}, \tag{2.38}$$

Equation (2.37) becomes

$$F(x',y',z',\ldots) = \alpha^n F(x,y,z,\ldots). \tag{2.39}$$

By differentiating with respect to α

$$\frac{\partial F}{\partial x'}\frac{\partial x'}{\partial \alpha} + \frac{\partial F}{\partial y'}\frac{\partial y'}{\partial \alpha} + \frac{\partial F}{\partial z'}\frac{\partial z'}{\partial \alpha} + \cdots = n\alpha^{n-1}F(x,y,z,\ldots). \tag{2.40}$$

Upon using Equation (2.38)

$$\frac{\partial x'}{\partial \alpha} = x, \qquad \frac{\partial y'}{\partial \alpha} = y, \qquad \frac{\partial z'}{\partial \alpha} = z, \text{ et cetera}; \tag{2.41}$$

and substituting these relations into Equation (2.40) we get

$$x\frac{\partial F}{\partial x} + y\frac{\partial F}{\partial y} + z\frac{\partial F}{\partial z} + \cdots = n\alpha^{n-1}F(x,y,z,\ldots). \tag{2.42}$$

If we now set $\alpha = 1$ in Equation (2.42) the result is

$$x\frac{\partial F}{\partial x} + y\frac{\partial F}{\partial y} + z\frac{\partial F}{\partial z} + \cdots = nF. \tag{2.43}$$

This is Euler's theorem on homogeneous functions.

It is appropriate to make some remark concerning the thermodynamic relevance of homogeneous functions. Consider a simple one-component fluid, such as water. If its temperature and pressure are held fixed, then its volume V will be directly proportional to its mass M. If V is expressed as a function $V(M)$ of M, then

$$V(\alpha M) = \alpha V(M) \tag{2.44}$$

for any α, since V is directly proportional to M. If M is increased by the factor α, so is V. We see by comparison of Equations (2.44) and (2.37) that V is a homogeneous function of the first degree in M.

Thermodynamic variables like V and M, which are proportional to the size of the system, are called *extensive* variables. Extensive variables are always homogeneous functions of one another in the first degree. For the record, it is worth pointing out that variables such as pressure, temperature, and density which are independent of M are referred to as *intensive* variables.

6. Constrained Extremals

As a final note in this chapter, we shall describe (again without rigor) the *Lagrange method of undetermined multipliers*. This is a technique for maximizing or minimizing a function $F(x,y,z,\ldots)$ when there are relations (constraints) among the variables $x,y,z,\ldots$. In order to be concrete, assume that there are N variables $x,y,z,\ldots$ and n relations among them ($n < N$) represented by the equations

$$\begin{aligned} \alpha_1(x,y,z,\ldots) &= 0, \\ \alpha_2(x,y,z,\ldots) &= 0, \\ &\vdots \\ \alpha_n(x,y,z,\ldots) &= 0. \end{aligned} \tag{2.45}$$

If these constraining relations did not exist, an extremum of F could be found by setting the differential of F equal to zero,

$$dF = \frac{\partial F}{\partial x}\,dx + \frac{\partial F}{\partial y}\,dy + \frac{\partial F}{\partial z}\,dz + \cdots = 0. \tag{2.46}$$

(Here the variables which are held constant for each partial derivative are not indicated. It is obvious which are held constant, and this leads to a simpler notation.)

Now if there were no relations between $x, y, z, \ldots$, Equation (2.46) could only be satisfied identically if each differential coefficient were itself zero,

$$\frac{\partial F}{\partial x} = \frac{\partial F}{\partial y} = \frac{\partial F}{\partial z} = \cdots = 0. \tag{2.47}$$

This follows from the independence of the variables $x, y, z, \ldots$. If values are chosen for $x, y, z, \ldots$ such that Equation (2.46) holds, one of the independent variables can always be varied, leaving the others fixed, so that the sum in (2.46) is no longer zero. The only exception occurs when Equation (2.47) is true. The N equations (2.47) are enough to determine values of $x, y, z, \ldots$ corresponding to the extremum.

When constraining conditions such as Equation (2.47) exist this procedure cannot be utilized, for the variables are no longer independent and (2.47) is no longer a necessary prerequisite for the validity of Equation (2.46). However, by use of a special device, we can bring the problem into a form such that conditions similar to (2.47) can be formulated. This is the method of undetermined multipliers.

From Equation (2.45) we obtain the total differentials of $\alpha_1, \alpha_2, \ldots \alpha_n$ and we get

$$\begin{aligned} d\alpha_1 &= \frac{\partial \alpha_1}{\partial x} dx + \frac{\partial \alpha_1}{\partial y} dy + \frac{\partial \alpha_1}{\partial z} dz + \cdots = 0, \\ d\alpha_1 &= \frac{\partial \alpha_2}{\partial x} dx + \frac{\partial \alpha_2}{\partial y} dy + \frac{\partial \alpha_2}{\partial z} dz + \cdots = 0, \\ &\vdots \\ d\alpha_n &= \frac{\partial \alpha_n}{\partial x} dx + \frac{\partial \alpha_n}{\partial y} dy + \frac{\partial \alpha_n}{\partial z} dz + \cdots = 0. \end{aligned} \tag{2.48}$$

Now each of these is multiplied by a parameter $\lambda_1, \lambda_2, \ldots \lambda_n$, respectively, the value of which for the moment remains unspecified. Thus,

$$\lambda_i\left[\frac{\partial \alpha_i}{\partial x} dx + \frac{\partial \alpha_i}{\partial y} dy + \frac{\partial \alpha_i}{\partial z} dz + \cdots\right] = 0, \qquad i = 1, 2, \ldots n. \tag{2.49}$$

If all the equations (2.49) are added to Equation (2.46) and the terms in dx, dy, dz, $\ldots$, are collected, we obtain

$$\begin{aligned} &\left[\lambda_1 \frac{\partial \alpha_1}{\partial x} + \lambda_2 \frac{\partial \alpha_2}{\partial x} + \cdots \lambda_n \frac{\partial \alpha_n}{\partial x} + \frac{\partial F}{\partial x}\right] dx \\ &\quad + \left[\lambda_1 \frac{\partial \alpha_1}{\partial y} + \lambda_2 \frac{\partial \alpha_2}{\partial y} + \cdots \lambda_n \frac{\partial \alpha_n}{\partial y} + \frac{\partial F}{\partial z}\right] dy \\ &\quad + \left[\lambda_1 \frac{\partial \alpha_1}{\partial z} + \lambda_2 \frac{\partial \alpha_2}{\partial z} + \cdots \lambda_n \frac{\partial \alpha_n}{\partial z} + \frac{\partial F}{\partial z}\right] dz + \cdots = 0. \end{aligned} \tag{2.50}$$

Now there are N variables and n relations between them, so that only $N - n$ of the variables are independent. Which of the N variables these are is immaterial. One has the liberty of choosing any $N - n$ of them. The remaining n variables are then dependent by default. In Equation (2.50) we can eliminate the coefficients of the differentials of these dependent variables by arbitrarily selecting the values of $\lambda_1, \lambda_2, \ldots \lambda_n$ so as to make these coefficients vanish. Since there are just as many λ's (undetermined multipliers) as coefficients of dependent variables (n of them), it is usually possible to accomplish this. Thus, if x is one of the dependent variables, the coefficient in Equation (2.50) corresponding to it is made to satisfy the following relation:

$$\lambda_1 \frac{\partial \alpha_1}{dx} + \lambda_2 \frac{\partial \alpha_2}{dx} + \cdots \lambda_n \frac{\partial \alpha_n}{dx} + \frac{\partial F}{\partial x} = 0. \tag{2.51}$$

The same holds for the coefficient of each of the n variables in the arbitrarily chosen dependent set.

As a result of this adjustment of the λ's, the only terms left in Equation (2.50) will be those corresponding to the $N - n$ *independent* variables. But now, this equation can only be true if, as in the case of Equation (2.47), each coefficient vanishes identically, for the differentials involved are all independent. Thus, for example, if z is one of the independent variables, we must have

$$\lambda_1 \frac{\partial \alpha_1}{\partial z} + \lambda_2 \frac{\partial \alpha_2}{\partial z} + \cdots \lambda_n \frac{\partial \alpha_n}{\partial z} + \frac{\partial F}{\partial z} = 0. \tag{2.52}$$

Since Equation (2.52) is exactly of the form of Equation (2.51), it follows that irrespective of whether a variable is chosen to be independent or dependent, the coefficient corresponding to it is zero. The specification of which of the variables are independent is therefore only a formal matter, and all the coefficients in Equation (2.50) can be equated to zero. Thus, we arrive at the counterpart of Equation (2.47). The N relations constituted by these vanishing coefficients together with the n equations of constraint (2.45) are enough to determine the N values of $x, y, z, \ldots$ and the n values of λ_1, λ_2, et cetera, corresponding to an extremum of F.

III

The First Law of Thermodynamics

1. Laws in Thermodynamics

Conventionally, thermodynamics is based upon three laws. However, it might be argued that there are really many others. This point of view becomes more comprehensible when it is realized (as we shall discuss later) that the first and second laws *postulate* that two measurable quantities (internal energy and entropy, respectively) are functions which depend only upon the thermodynamic variables of state. In fact, they may themselves serve as variables of state since the functional relationships may be used to eliminate some previously chosen state variable. Thus, functions of state may be used as state variables and vice versa.

As a result of this, the assumption that any property, for example volume, is an independent variable of the thermodynamic system is equivalent to the assumption that it is a function of state. In this sense, the process of assuming that a given property may be employed as an independent variable is logically equivalent to the process embodied in, say, the first law of thermodynamics in which *internal energy* is assumed to be a function of state. Thus, when we say that volume is a state variable we are introducing a law.

Unlike volume, however, *internal energy* and *entropy* are not deeply rooted in intuition, and a considerable amount of discussion is required before their function-of-state characters can be rendered plausible. It is for this reason more than any other that the assumptions concerning them are dignified by the appellation "law."

An important assumption, not ordinarily counted among the laws of thermodynamics concerns the existence of the variable *temperature*, a quantity which is closer to intuition than internal energy or entropy but more obscure than *volume*. Sometimes the assumption of the existence of temperature is referred to as the *zeroth law of thermodynamics* to emphasize that it is anterior to the usual three. This chapter will be concerned mainly with the first law of thermodynamics, but before we can proceed with the exposition it is necessary to understand the concept of temperature.

2. Temperature

In defining temperature we must avoid all recourse to the idea of *heat*. This is because heat in itself has no independent existence. In the early days of thermodynamics, heat like electricity was regarded as a fluid which could be contained within a body in definite amounts. As a result, it was thought that the amount of heat stored within a body might play the role of an independent thermodynamic variable. As the discipline evolved, it became clear that no such fluid existed, and that what had been called heat was in fact a form of energy which could be transformed into other kinds of energy; for example, into mechanical *work*. Since energy in the form of work could be measured by techniques available at the time, it was work which assumed the primary position with heat being relegated to a secondary place in the hierarchy of kinds of energy.

Although heat is not conserved, it does have a meaning and we shall take steps to define this meaning carefully. All definitions, however, will be made in terms of the purely mechanical notion of work. By proceeding carefully, we shall be able to define temperature using nothing more than mechanical ideas.

The development will require the introduction of certain idealized walls which can enclose, but are not part of, the substances to be dealt with. These will be of two sorts, adiabatic and diathermic. In anticipation of the definition of heat, we can say that adiabatic walls will not permit the flow of heat whereas diathermic walls will. However, as the concept of heat has not yet been introduced, we must seek means of definition which make no mention of heat.

Consider a simple substance, perhaps a fluid, whose thermodynamic variables (once its mass has been fixed) are two in number. These we can choose to be the pressure p and the volume V, both of which are purely mechanical notions. We enclose this system by a wall impermeable to matter and assume that there are no fields present—electromagnetic, gravitational, et cetera—or at least no fields which can be varied. Under this condition, work can only be performed on the system through the movement of part of its wall.

The *adiabatic* wall is now defined by the following properties:

> If the above system is in equilibrium and is enclosed by an adiabatic wall, then it is possible to disturb it (excluding force fields) only through the movement of a part of the wall, but by no other external action.

In anticipation of the heat concept, let us say that such a wall does not permit changes of state which are due to heat, but only those which are brought about by the expenditure of mechanical work which, in the absence of external force fields, can only be accomplished by the movement of parts of the wall (stirring, compression, and so forth).

Some elaboration of this point is desirable. Consider the performance of work on the system which may be achieved by stirring. By anticipating the heat concept once again, we can expect that the change of state which attends stirring will be manifested by a change in temperature (anticipating the temperature concept). From the mechanistic point of view the heating due to stirring will be the result of friction, and the mechanical work of stirring will be performed against the forces of friction. Now friction may occur between the paddle and the fluid being stirred, or it may occur internally within the fluid itself. In the former case, the paddle would have to be very ideal indeed, for it must be able to heat the fluid by frictional means without becoming hot itself. Otherwise, having accepted heat, and by definition *not being part of the system*, it will constitute a portion of the wall which is not adiabatic.

Stirring does not necessarily have to be performed by a paddle. The walls might be flexible, for example, and by deforming them one might be able to disturb the fluid sufficiently so that its motion generates internal friction. By stirring we shall generally mean work performed against frictional forces. One might move the wall without stirring, that is, without generating frictional forces. This process is compression or expansion, and we have already discussed the *volume* work associated with it in Chapter 1, Section 6.

The use of abstractions such as highly idealized walls is a feature which permeates all of thermodynamics. Although these procedures leave something to be desired, no simple method exists by means of which they can be eliminated. Carathéodory's approach, which will be discussed in Chapter 4, comes closer to accomplishing this but does not succeed entirely.

The *diathermic* wall is defined as one which does not conform to the above criterion. But a more useful definition is the following:

> If two simple substances otherwise adiabatically enclosed (by walls impermeable to matter) are separated from each other by a diathermic wall (impermeable to matter), then they cannot be at equilibrium at arbitrary values of their variables of state, p_1, V_1 and p_2, V_2. For this to be the case there must be a certain relation between the four variables. That is,
>
> $$\beta(p_1, V_1, p_2, V_2) = 0. \tag{3.1}$$
>
> Such an equation is the expression of thermal contact.

We are all familiar with the purely subjective feelings of hot and cold. Although we are accustomed to thinking of bodies with higher temperatures being hotter and those with lower temperatures being colder, we cannot appeal to these sensations for the definition of temperature and certainly not for the definition of heat. On the other hand, Equation (3.1) can be used very expeditiously for this purpose. The concept of temperature is based on the fact of experience that two bodies in thermal equilibrium with a third are also in equilibrium with each other. If one employs Equation (3.1) as an expression of thermal contact, the last statement reads as follows:

The equations

$$\beta_1(p_2,V_2,p_3,V_3) = 0, \qquad \beta_2(p_1,V_1,p_3,V_3) = 0 \tag{3.2}$$

require the equation

$$\beta_3(p_1,V_1,p_2,V_2) = 0, \tag{3.3}$$

or by permutation of indices, from the validity of any two of these relations follows the validity of the third.

From this assumption we may develop a very convenient relationship. Solve the first of Equations (3.2) for p_3 and obtain

$$p_3 = \varphi_1(p_2,V_2,V_3). \tag{3.4}$$

Similarly, solve the second of Equations (3.2) for p_3,

$$p_3 = \varphi_2(p_1,V_1,V_3). \tag{3.5}$$

By eliminating p_3 between Equations (3.4) and (3.5), we obtain

$$\varphi_1(p_2,V_2,V_3) = \varphi_2(p_1,V_1,V_3). \tag{3.6}$$

But Equations (3.4) and (3.5) are expressions of the fact that bodies 1 and 2 are separately in equilibrium with body 3 (since they are derived from Equation (3.2)); therefore, according to the above postulate, Equation (3.3) must hold between the variables p_1,V_1,p_2,V_2. If this is to be true, V_3 in Equation (3.6) must be contained in φ_1 and φ_2 in such a manner that it cancels out of the equation. The most general forms which would permit this to happen are the following:

$$\varphi_1(p_2,V_2,V_3) = f_2(p_2,V_2)h(V_3) + q(V_3), \tag{3.7}$$

$$\varphi_2(p_1,V_1,V_3) = f_1(p_1,V_1)h(V_3) + q(V_3). \tag{3.8}$$

Substitution of these into Equation (3.6) yields

$$f_1(p_1,V_1) = f_2(p_2,V_2). \tag{3.9}$$

A similar argument in which either of Equations (3.2) is regarded as a consequence of the other Equations, (3.2) and (3.3), leads to the results

$$f_1(p_1,V_1) = f_3(p_3,V_3) \tag{3.10}$$

and

$$f_2(p_2,V_2) = f_3(p_3,V_3). \tag{3.11}$$

Thus, one can always write the condition for equilibrium between two substances in the form (3.9). It is of utmost importance to realize that the important feature of Equations (3.9), (3.10), and (3.11) is the fact that each function f_i depends only upon the variables p_i and V_i of the ith substance; that is, the consequence of Equations (3.2) and (3.3) is not merely some relation of little value like $f_1(p_1,V_2) = f_2(p_2,V_1)$. Either substance can then be used as a *thermometer*, for we can employ the numerical value of function f_1 or f_2 as an *empirical temperature* θ. Thus,

$$f_2(p_2,V_2) = \theta. \tag{3.12}$$

Then the condition of equilibrium in the form (3.9) asserts that the first substance is in equilibrium with the second (the thermometer) if a certain relation,

$$f_1(p_1,V_1) = \theta, \tag{3.13}$$

exists between variables of state p_1, V_1 and the empirical temperature θ. This equation is called the *equation of state* for the substance, and we shall have more to say about it in Chapter 5. The associated p-V curves are called *isotherms*. Any other function of θ could, with equal validity, have been chosen as the empirical temperature. The isotherms always remain the same. This choice is limited only from the practical point of view. One chooses only those systems for thermometric substances which do not contain two phases, for example fluids below the point of condensation or in the gaseous state. Otherwise, the single valuedness of the thermometric reading may be compromised.

3. Temperature Scales

For a detailed discussion of temperature scales, the reader is referred to the many excellent texts in the field, and we shall have more to say on the subject in Chapters 4 and 5. For the moment, we will limit our comments to a few basic features of the problem. It is absolutely necessary to emphasize the extremely arbitrary manner in which a temperature scale is chosen. There is a preference for the use of a gas as a thermometric substance, but this preference is justified only by the fact that such a thermometer gives readings which are almost independent of the choice of the gas over a wide range of temperature. This is due to the fact that for all gases in states of high dilution, experiments show that the isotherms can be represented by the hyperbola $pV =$ constant. That one chooses just this product pV as the temperature of the gas and not some function of it, such as $(pV)^2$ or $\sqrt{pV}$, cannot be justified logically from the standpoint of theory. It is founded on the argument of simplicity. Later, when we have discussed the second law of thermodynamics, it will be possible to define a *thermodynamic* scale of temperature based on the efficiency of a thermal cycle and not on a substance. This scale will bear an especially simple relation to that of the gas thermometer. However, this in no way reduces the arbitrariness possible in the choice of any temperature scale. The use of the word *empirical* in the definition of θ is meant to emphasize this situation.

The assumption that the relations (3.2) and (3.3), or similar ones, are general for *all* substances is essentially the assumption of the existence of a temperature. As we have indicated, this assumption has been called the *zeroth law of thermodynamics.*

After the temperature has been defined, one can select either (p_1,V_1), (p_1,θ), or (V_1,θ) as the variables of state.

4. Adiabatic Work

We now turn to the task of introducing the first law of thermodynamics. It is in part an expression of the general principle of the conservation of energy, but in addition contains other features peculiar to thermodynamics. Along the way, we shall develop an appropriate definition for the term *heat.* It has been mentioned earlier that heat has no independent existence of its own. It cannot be said of a body that it contains a given amount of heat, although it is common in coloquial parlance to do so. In fact, heat is a term relevant only to processes involving the *transfer* of energy, that is, to the passage of energy from one system to another. This will become clear shortly.

As a result of much experimentation, the following has been demonstrated with a high degree of certainty. Suppose a system of fixed mass (let it be a simple substance like water) is enclosed by adiabatic walls (adiabatic in the sense we have defined the word). This state (call it state 1) can be represented as a point in the macroscopic state space of the system. If now a certain amount of work is performed on the system, the state is changed to a new one (state 2). This can also be represented as a point in state space. Now the important observation which has been made is the following.

The *amount* of work expended in passing to state 2, adiabatically, seems in no way to depend upon *how* the work is performed. For example, part of the work might result from compression and the rest from stirring (see Section 2). The same *total* quantity of work will have been performed regardless of the proportion of each, provided that the process always takes place between the same states 1 and 2.

In this connection, it is worthwhile to note that the inverse of this statement is not observed to hold generally. There are systems in which, for any given state 1, there are a plurality of states—state 2, state 2′, state 2″, et cetera—to which the system may be displaced adiabatically such that the amount of work performed is the same for each displacement. Thus, the performance of a definite amount of adiabatic work does not necessarily guarantee arrival at a unique end state; whereas, arrival at a unique end state, starting from a unique initial state, seems to guarantee the performance of a definite amount of work.

The above considerations suggest the possibility of defining a quantity characteristic of each state whose difference between two states can be measured by the amount of work which must be performed under adiabatic conditions to induce a transition between these states. Thus, if we denote the quantity by U, and its values in states 1 and 2 by U_1, and U_2, respectively, and if $-\Delta w_a$ is the amount of adiabatic work performed upon the system,

$$-\Delta w_a = U_2 - U_1. \tag{3.14}$$

U must be a function of state, depending on the variables of state, and nothing more. Otherwise, $-\Delta w_a$ would depend on more than just the initial and final states, contrary to experience. Equation (3.14) establishes a means by which U may be measured in any state to within an additive constant. By starting from some arbitrary reference state in which U has the value U_0, one contrives to bring the system into any other state by adiabatic means and measures the work performed. In an arbitrary state then,

$$U = -\Delta w_a^0 + U_0, \tag{3.15}$$

where $-\Delta w_a^0$ is the work performed, starting from the reference state. It must be emphasized that the existence in all cases of a function of state with the properties of U is in the final analysis pure assumption. It is an eminently reasonable one, based on our experience, but still an assumption.

At this juncture one may raise the question concerning whether or not it is possible to reach all states of the system by adiabatic means, starting from some arbitrary reference state. If there are states which are inaccessible, then the usefulness of Equation (3.15) as a means of specifying U becomes limited. Experience seems to provide the following answer to this question. There *are* states which are inaccessible by adiabatic means, starting from an arbitrary reference state. On the other hand, in these cases it seems possible always to overcome the difficulty in one of two ways:

1. It may be possible to *start* from the inaccessible state and transfer the system to the reference state by adiabatic means. In other words, if the adiabatic path between the reference state and an adjoining state cannot be traversed in one direction, it may be traversable in the other.
2. It may not be possible to transfer the system adiabatically from the reference state to the final state or vice versa, but then it always seems possible to traverse the path from or to *both* the reference and final states

to or from a third state. This situation may be illustrated diagrammatically if we denote the reference state by A and the final state by C, while the third state is symbolized by B. Then four adiabatic processes are conceivable:

$$A \rightarrow B \rightarrow C \qquad (1)$$

$$A \leftarrow B \leftarrow C \qquad (2)$$

$$A \rightarrow B \leftarrow C \qquad (3)$$

$$A \leftarrow B \rightarrow C \qquad (4)$$

Processes (1) and (2) are of no interest because they represent the *direct* transfer of the system from A to C or vice versa, state B merely serving as an intermediate one. However, processes (3) and (4) provide no path between A and C, but they do connect A and C to B.

Insofar as the application of Equation (3.15) is concerned, the fact that the final state is inaccessible by adiabatic means from the initial one presents no problem, for $-\Delta w_a^0$ may be measured just as easily by performing the inverse process (when possible) and measuring $+\Delta w_a^0$. When the inverse process is not possible, the initial and final states may be connected by a third state (as in processes (3) and (4) above), and the two works corresponding to the connection with this third state may be measured. In the latter case, one measures the change in U, proceeding from the reference state to the third state and its change between the third state and the final state. Thus, in effect, the total change in passing from the reference to the final state may be measured.

It is worthwhile to illuminate, by consideration of an example, the case in which a direct path (in at least one direction) exists between the reference and the final state. Consider a one-component gas enclosed in a cylinder and bounded by a movable piston. Let the walls and piston be adiabatic so that any change of volume must take place without heat being exhanged with the surroundings. The system can only exchange energy with its surroundings (unless a stirrer is inserted through an adiabatic seal) through the performance of work on the piston. The variables of state may be chosen to be temperature and volume. It is a matter of experience that the changes of volume which this system can undergo may be classified into four distinct kinds, only three of which are possible. For convenience, we shall designate these four by the letters a, b, c, and d.

In change a, the system expands with a decrease in temperature. If the volumes and temperatures of the initial and final states are designated by V_0, T_0 and V_1, T_1, respectively, this process can be represented symbolically by the following inequalities:

$$\begin{aligned} T_1 &< T_0, \\ V_1 &> V_0. \end{aligned} \qquad (3.16a)$$

In change b, the system is compressed and its temperature rises. This process can be represented symbolically by

$$T_1 > T_0, \quad V_1 < V_0. \tag{3.16b}$$

In change c, the system is expanded and its temperature increases. The corresponding inequalities are

$$T_1 > T_0, \quad V_1 > V_0. \tag{3.16c}$$

Some discussion of change c is desirable. Ordinarily when a gas is expanded adiabatically its temperature drops. Thus, it will not be possible to carry out change c by expansion alone, for in that process the gas performs work on its environment through the medium of the piston, and its temperature is reduced. On the other hand, after expansion to the proper volume has been achieved, it is possible to insert a stirrer into the gas through an adiabatic seal and perform work while the volume remains constant. This will lead to an increase in gas temperature. Through sufficient stirring, therefore, it is possible to bring the gas in its expanded state to a temperature T_1 greater than T_0. Thus, process c is possible; but it must be done in two stages utilizing two different kinds of work, expansion work and stirring work.

We now come to change d. This is a process in which the gas is compressed while its temperature drops. The appropriate inequalities are

$$T_1 < T_0 \quad V_1 < V_0. \tag{3.16d}$$

We notice that this process is impossible. Adiabatic compression of the gas leads invariably to an increase in gas temperature. Nor can we reduce the temperature through the performance of additional work. Thus, we have arrived at what seems to be an impossible change.

Insofar as the application of Equation (3.15) is concerned, however, there is no cause for alarm, for $-\Delta w_a^0$ can be measured just as easily by performing the inverse process and measuring its negative. That the inverse process is a possible one becomes evident immediately when it is realized that the inverse is nothing more than change c which has already been classified as possible.

It is of interest to summarize the inequalities contained in Equations (3.16a, b, c, d) by referring them to the macroscopic state space of the simple gas in question. This is illustrated in Figure 3.1. Here volume is the ordinate

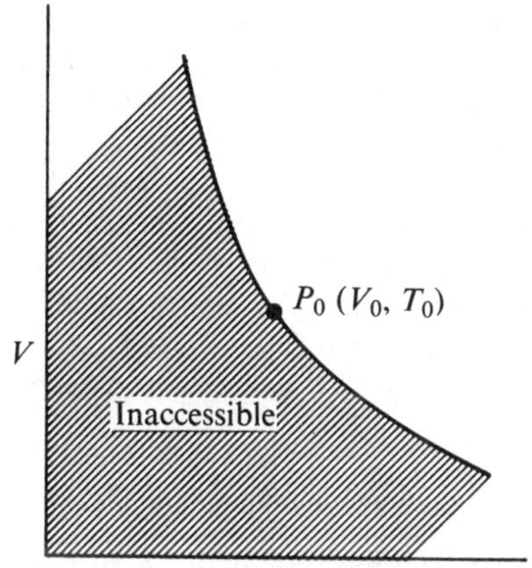

FIGURE 3.1. *Volume-temperature diagram of a simple substance illustrating the region (shaded) which is inaccessible by adiabatic means from the reference state P_0.*

and temperature is the abscissa. The point P_0 represents the initial state, and the shaded region denotes that region of state space which is inaccessible by adiabatic means starting from point P_0. The reader may verify that the above-mentioned inequalities are consistent with the figure.

The precise shape of the boundary curve separating the inaccessible from the accessible region in Figure 3.1 depends upon the specific substance involved. The discussion of cases *a*, *b*, *c*, and *d* above, however, does indicate that the slope, $\frac{dV}{dT}$, at P_0 must always be negative.

5. Internal Energy, the First Law, and Heat

The quantity U obviously has the dimensions of energy. It has been assigned the name *internal energy*. We are now in a position to state the first law of thermodynamics:

> The first law of thermodynamics postulates the existence of a function of state U called the internal energy and specifies a method for measuring it to within a constant. This method, which is embodied in the relationship expressed by Equation (3.15) involves the determination of the adiabatic work expended in passing from some reference state of the system to the state in question.

The undetermined constant in Equation (3.15) causes no difficulty since (as will become clear as the subject unfolds) for thermodynamic purposes it is only necessary to know the *difference* in internal energy between two states, and in taking this difference the constant cancels.

Suppose U_2 and U_1 have been measured in this fashion for two thermodynamic states, 1 and 2. If now the system in brought from state 1 to state 2 under *nonadiabatic* conditions and the work $-\Delta w$ performed during this change is measured, it is a matter of experience that, depending on states 1 and 2, Δw is not always equal to $U_2 - U_1$ as is $-\Delta w_a$ when the change is brought about adiabatically. We denote the difference between $U_2 - U_1$ and $-\Delta w$ by the symbol Δq; thus,

$$U_2 - U_1 + \Delta w = \Delta q. \tag{3.17}$$

Δq is by definition the *heat* absorbed by the system in passing from state 1 to 2.

This is the best meaning which can be attached to heat. It is simply an adjustment term to balance the energy transfer to the system during a nonadiabatic change. As we have defined heat, it is a completely secondary quantity in the sense that it is only measured in terms of work. It is assumed that U measures an energy content in the system (U itself has been measured in terms of work). In a nonadiabatic process, when the energy content of the system seems to change by more than can be accounted for by the work performed on the system, it is possible to say the additional energy has passed into the system in the form of heat. Its quantitative measure is given in Equation (3.17).

Notice that this in no way implies that the system contains a given amount of heat; energy, yes, but not heat. Heat is simply a discrepancy term measurable during nonadiabatic changes.

It should be noted (see Chapter 1, Section 6) that $-\Delta w$ refers to work performed *on* the system. Thus, Δw in Equation (3.17) represents the work performed *by* the system.

For an infinitesimal change, a differential process, Equation (3.17) may be written in the form

$$dU = Dq - Dw. \tag{3.18}$$

In conformance with the conventions of Chapter 2, the symbol d is employed as a prefix for U in Equation (3.18). Since U is a function of state, it has a definite value for every point in the macroscopic state space. Thus, dU is an exact or complete differential. On the other hand, the symbols Dq and Dw with their large D's refer to quantities which are not necessarily differentials of functions of state. A system at a given point in state space does not have a given amount of heat or work. Neither heat nor work is a function of state.

Equation (3.18) if often regarded as a form of the first law of thermodynamics. The reader should be reminded at this point that since all our arguments (on adiabatic work) have been developed in connection with systems of fixed mass, Equation (3.18) is only valid for such systems. A system of fixed mass which cannot exchange matter with its surroundings is termed a *closed* system in thermodynamics. In contrast, a system which can exchange matter with its surroundings is said to be *open*. Closed systems are to be distinguished from *isolated* systems which are prohibited from exchanging energy as well as matter with their surroundings.

6. Heat Capacity, Enthalpy, and Heat of Change

Our discussion has emphasized the fact that the first law is concerned with postulating the existence of a function of state, namely the internal energy. In Section 1, some pains were taken to point out that functions of state and variables of state could be used interchangeably, and that the zeroth law, which assumes the existence of a temperature, is therefore also concerned with postulating the existence of a function of state.

It will become evident, as we develop our theme, that the power of the thermodynamic method is related to the number of functions of state which we are able to recognize. In this connection it is important to distinguish between what might be called *primary* and *derived* functions of state. Primary functions are those, such as internal energy, whose existence must be postulated whereas derived functions are combinations of primary ones.† The discovery of a primary function is of paramount importance in thermodynamics, but derived functions are also useful; and we shall discuss a few of these in the present section.

A useful derived function is C_v, *the heat capacity at constant volume*. This is defined by the relation

$$C_v = \left[\frac{\partial U}{\partial T}\right]_V. \tag{3.19}$$

T is the ideal gas or Kelvin temperature, to be defined later. C_v could be defined quite generally in terms of the empirical temperature θ; but since we shall use it only in conjunction with T, confusion may be avoided by defining it in terms of T immediately.

Another useful derived quantity is the enthalpy H,

$$H = U + pV. \tag{3.20}$$

† Actually, it is possible to postulate the existence of some of those functions of state which we have called "derived" and then to derive those which have been called "primary." The selection of a set of primary functions is a matter of convenience.

If a system is set up so that it may perform only volume work during a change of state which carries it between initial and final states characterized by the same value of pressure p, it is usually assumed that the heat Dq absorbed during the change is measured by dH; a fact which has led some to call H the *heat content*. Although in most instances this represents an excellent approximation, there are some limitations to this point of view which are not usually discussed when proofs are offered.

In the first place, Dw in Equation (3.18) may be replaced, according to Equation (1.1), by $p\,dV$ so that

$$Dq = dU + p\,dV. \tag{3.21}$$

In this equation p stands for a hydrostatic pressure maintained (by some mechanism) in the medium surrounding the system. If the change occurs with sufficient velocity, there is no assurance that a pressure p can even be defined for the system itself (since very general stresses may be set up), but Equation (3.21) remains valid so long as the surrounding medium continues to behave hydrostatically. For example, suppose that the change in question is the occurrence of a chemical reaction and that dV is the increment in volume which attends this reaction. This may take place very rapidly, but the surrounding medium may relax so fast that it maintains a pressure p (to a high degree of approximation) during the reaction. Then Equation (3.21) will apply. This is, in fact, the usual situation.

Now the p in Equation (3.20) is the pressure of the system. If a thermodynamic change occurs between states having the same value of p, differentiation of Equation (3.20) will yield

$$dH = dU + p\,dV, \tag{3.22}$$

where the term $V\,dp$ is necessarily zero and is therefore not shown. Comparison of Equation (3.22) with Equation (3.21) reveals that Dq is given by dH. When the surrounding medium does not exert a uniform hydrostatic pressure during the change, Equation (3.21) is invalid and the comparison cannot be made.

Another useful derived quantity is C_p, the heat capacity at constant pressure.

$$C_p = \left[\frac{\partial H}{\partial T}\right]_p. \tag{3.23}$$

This formula gives recognition to the fact that the heat absorbed at constant pressure is dH rather than dU, the latter being the heat absorbed at constant volume.

7. Phases

Certain derived quantities are defined with respect to single phases, and it is appropriate to supply a reasonable definition of a *phase*:

> A phase is a part of a system, physically distinct, macroscopically homogeneous, and of fixed or variable composition. It is mechanically separable from the rest of the system.

How well these criteria are satisfied depends again on the discretion of the experimenter. For the phase to be macroscopically homogeneous, for example, it should have a smooth and uniform texture such as is possessed by a solution of salt in water, or pure solid sugar, or clear air devoid of fog or dust. Liquid water, ice, and water vapor represent different phases of the same substance which, under certain conditions, can coexist in contact with one another. The questions of physical distinctness and mechanical separability arise when two or more such phases are in contact. Consider, for example, liquid in contact with vapor. Inside the bulk liquid the situation is uniform and homogeneous in accord with our definition of a phase, and the same can be said for the interior of the bulk vapor. On the other hand, it is well-known that on a microscopic scale the boundary between the liquid and the vapor is not mathematically abrupt. That is, if we were to plot the density as a function of distance through the interface, the resulting curve would not show a mathematical discontinuity at some point where the density would jump from its value characteristic of the vapor to that of the liquid. Rather, over a distance of some 10–20 angstroms there would be a rapid but continuous change of density. As a result, the situation in the interfacial region is nonhomogeneous and does not fulfill the phase requirement enunciated above. Thus, if the system is examined on too fine a scale it would not be possible to say where one phase ends and the other begins. Under this circumstance, it would be difficult to claim with perfect assurance that the two phases were physically distinct or mechanically separable. On the other hand, in most applications our measurements will be so gross that the amount of material in the inhomogeneous interfacial region will be negligible and no real problem will arise.

It is clear that a phase is a chemical solution in the classic sense. It may consist of k components. In chemical parlance the numbers of moles of each can be designated by $n_1, n_2, \ldots, n_k$. These mole numbers can comprise part of the set of independent variables of the system. For example, a suitable set of independent variables might be temperature, pressure, and the mole numbers.

8. Intensive and Extensive Quantities

In Chapter 2 a distinction was made between the intensive and extensive properties of a thermodynamic system. The former were quantities like temperature, pressure, and density which, in a homogeneous phase with uniform properties, could be measured at different points within the phase; and in the absence of external fields, could be independent of the position of measurement.

Extensive properties are quantities, like volume, which are related to the extent of the system in a linear manner. Thus, doubling the mass of a system while maintaining its intensive properties constant, doubles its volume.

The internal energy U is an extensive quantity. Until now we have made no effort to prove this, and some discussion is necessary. Consider a system which can only exchange volume work with its surroundings. If the internal energy of this system is measured by an adiabatic process in accordance with Equations (1.1) and (3.15), we have

$$dU = p\,dV, \tag{3.24}$$

provided the process is conducted reversibly so that p is the equilibrium pressure. Suppose we are in possession of another system identical in all respects to the one involved in Equation (3.24). Then it can be made to undergo the same change. If primes are used to indicate the quantities referring to this system, we have

$$dU' = p\,dV', \tag{3.25}$$

where p has been used in place of p' because the pressure in both systems is the same by definition.

Now it is clear that if the two identical systems were joined physically, not only would the composite system have the same intensive property p, but the total volume V_t would be exactly $V + V'$, where V and V' are the individual volumes before the systems are merged. Thus, the differential of internal energy for the total system is

$$dU_t = p\,dV_t = p(dV + dV') = dU + dU' = d(U + U'). \tag{3.26}$$

From this it is apparent that

$$U_t = U + U' = 2U. \tag{3.27}$$

Thus, the internal energy of the system with double mass is itself double since $U = U'$. In the case under discussion, internal energy is therefore an extensive quantity. We have advanced the argument by considering a system of double size. It is evident that the same result can be demonstrated for a system of arbitrary size—increasing the mass by a given factor increases the internal energy by the same factor, provided that the intensive variables remain fixed.

This kind of proof must be performed anew each time the question arises, for example, when the system cannot be brought adiabatically from one state to another by the mere performance of volume work. The result is always the same, however; if the intensive variables have uniform values throughout the system, then U is an extensive quantity. Clearly, functions of state like the enthalpy (Equation (3.20)) which are linear combinations of extensive quantities are themselves extensive.

9. Euler's Theorem and Partial Molar Quantities

If the mole number of each component of a given solution (phase) is changed by the same factor α, then the mass of the phase will also be changed by the factor α. In Chapter 2 the point was made that extensive variables are homogeneous functions in the first degree of the mass of the system; the intensive variables being held constant. From what has just been said, it follows that they are also homogeneous functions in the first degree of the mole numbers.

Consider the volume of the phase. From Euler's theorem (2.43) we have

$$V = n_1\left[\frac{\partial V}{\partial n_1}\right]_{T,p,n_2,n_3,\ldots} + n_2\left[\frac{\partial V}{\partial n_2}\right]_{T,p,n_1,n_3,\ldots} + \cdots, \tag{3.28}$$

where we have assumed that a suitable set of independent thermodynamic variables consists of the two intensive quantities T, p, and the mole numbers. Now it is customary to call the partial derivatives appearing in Equation (3.28) *partial molar volumes* and to denote them by symbols with bars. Thus,

$$\bar{V}_1 = \left[\frac{\partial V}{\partial n_1}\right]_{T,p,n_2,n_3,\ldots} \tag{3.29}$$

is the partial molar volume of species 1. In fact, this partial derivative with respect to n_i of any extensive quantity, is a partial molar quantity.

With this notation, Equation (3.28) can be rewritten as

$$V = n_1\bar{V}_1 + n_2\bar{V}_2 + \cdots n_k\bar{V}_k. \tag{3.30}$$

From Equation (3.30) it appears, superficially, as though the solution were assembled from components having the volumes $\bar{V}_1$, $\bar{V}_2$, et cetera, per mole in the separated state. Actually, $\bar{V}_i$ bears no necessary relation to V_i^0, the volume per mole of pure species i, unless the solution consists of only one component and is therefore pure i. In that case,

$$\begin{aligned} V &= n_i\left(\frac{V}{n_i}\right) = n_iV_i^0, \\ \bar{V}_i &= \left[\frac{\partial V}{\partial n_i}\right]_{T,p} = V_i^0 = \frac{V}{n_i}. \end{aligned} \tag{3.31}$$

Here we have omitted the subscripts such as n_2, n_3, et cetera, which appear in Equation (3.29) and which indicate that all the mole numbers, except n_i, are held constant. It is, however, to be understood that this is the case. No confusion should result, and for notational simplicity this practice will be followed from now on.

Our method of viewing $\bar{V}_i$ is the following. Imagine a solution of infinite extent having the same values for its intensive variables as the solution in question. To this infinite solution we add one mole of species i, keeping the temperature and pressure constant. During this process, the volume of the solution will probably change. This change is measured by $\bar{V}_i$. Alternatively, one may add an infinitesimal quantity dn_i of species i to a finite solution. This will occasion an infinitesimal change dV in volume. The ratio of dV to dn_i then measures $\bar{V}_i$. In either case, the change in volume is shared jointly by the species already in the solution as well as by the material being added.

It is now possible to derive an interesting relationship between changes in temperature and pressure, and the associated changes in partial molar volumes of the component species. Changes in composition may be occurring, but the differentials of the mole numbers do not appear explicitly in the relation. If we assume that temperature, pressure, and the mole numbers are the independent variables, then the total differential of the volume with respect to these variables is

$$dV = \left[\frac{\partial V}{\partial T}\right]_{p,n_1,n_2,\ldots} dT + \left[\frac{\partial V}{\partial p}\right]_{T,n_1,n_2,\ldots} dp + \sum_i \bar{V}_i\, dn_i. \tag{3.32}$$

Now obtain the total differential of the volume by differentiating Equation (3.30). The result is

$$dV = \sum_i n_i\, d\bar{V}_i + \sum_i \bar{V}_i\, dn_i. \tag{3.33}$$

Comparison of Equations (3.32) and (3.33) reveals that

$$\sum n_i\, d\bar{V}_i = \left[\frac{\partial V}{\partial T}\right]_{p,n_1,n_2,\ldots} dT + \left[\frac{\partial V}{\partial p}\right]_{T,n_1,n_2,\ldots} dp. \tag{3.34}$$

This is the relation we have been seeking.

This equation furnishes an example of the service which thermodynamics offers, namely that of establishing interconnections between the measurable thermodynamic properties of a given system. The technique of derivation involves partial differentiation of the various state functions. It is clear that the richer the field of *independent* state functions the larger will be the harvest of relations of the type (3.34).

If pressure and temperature are held constant, Equation (3.34) simplifies to

$$\sum_i n_i\, d\bar{V}_i = 0. \tag{3.35}$$

Since temperature and pressure are held constant, the process to which Equation (3.35) corresponds can only involve changes in composition. It might, for example, involve a chemical reaction. From the discussion in Section 6, the heat Dq absorbed by the system during this reaction may be represented (within the limitations cited in Section 6) by dH because the pressure is held constant. The analogue of Equation (3.30) for H is

$$H = \sum_i n_i \bar{H}_i, \tag{3.36}$$

and by differentiation,

$$dH = \sum_i n_i \, d\bar{H}_i + \sum_i \bar{H}_i \, dn_i. \tag{3.37}$$

For constant temperature and pressure, the analogue of Equation (3.35) holds so that Equation (3.37) reduces to

$$(dH)_{T,p} = Dq = \sum_i \bar{H}_i \, dn_i. \tag{3.38}$$

Dq refers to the heat *absorbed* by the system, whereas the heat of chemical reaction (denoted by DQ) is the heat *evolved* by the system. Thus,

$$DQ = -Dq = -\sum_i \bar{H}_i \, dn_i. \tag{3.39}$$

This equation forms the basis of the subject of thermochemistry.

§ IV

The Second Law and Entropy

1. General Remarks

The *second law* of thermodynamics may be stated in the following form:

> For any thermodynamic system there exists a function of state which we shall call the *entropy* and denote by the symbol S. The method for measuring entropy is based upon the fact that corresponding to any change of state for the system conducted *reversibly* (quasistatically)
>
> $$dS = \frac{Dq}{g(\theta)}, \tag{4.1}$$
>
> where Dq is the heat absorbed by the system during the reversible change and $g(\theta)$ is a definite but as yet unspecified function of the empirical temperature θ alone. Furthermore, for any change which can in fact take place in an *isolated* system,
>
> $$\Delta S > 0 \tag{4.2}$$
>
> (where the Δ stands for a finite change).

According to the definition presented in Chapter 3, Section 6, an *isolated* system is one which can exchange neither matter nor energy with its environment. This limitation may be achieved by surrounding the system with rigid adiabatic walls which are impermeable to matter.

The above statement of the second law is formal and abstract. Nevertheless, it is equivalent to other statements which may seem more deeply rooted in intuition. Notice that in many respects the second law is similar to the first. Like the first, which asserts the existence of a function of state, *the internal energy*, the second law asserts the existence of another function of state, the *entropy*. Furthermore, just as the first law prescribes a method for the determination of the internal energy (through the measurement of work performed adiabatically), the second law prescribes a method for the measurement of entropy through the application of Equation (4.1). Thus, the entropy difference between two states 1 and 2 is

$$S_2 - S_1 = \int_1^2 \frac{Dq}{g(\theta)}, \tag{4.3}$$

where the integral is taken over a reversible or quasistatic path (a path which can be plotted in state space).

However, the inequality contained in Equation (4.2) is foreign to the *first* law. This part of the second law has valuable consequences.

The function $g(\theta)$ is not perfectly arbitrary but (except for a scale factor) is related in a perfectly definite way to other measurable physical quantities. However, the precise form of $g(\theta)$ need not be specified in the statement of the second law since, as we shall show below, it can be derived as soon as the method for measuring temperature is defined.

Many authors have sought to make the second law seem reasonable by appealing to the facts of experience. From the point of view of the pedagogue, the formal statement advanced above leaves much to be desired. Any attempt to base the second law in intuition, however, should make it clear from the start that the law *cannot be proved.* Like the first law, it is a postulate, albeit an eminently *reasonable* one.

The teaching of the second law presents an unending challenge to the pedagogue. Energy seems more real to the student than entropy for which his intuitive grasp comes less from everyday experience. It is also true that the student will have enjoyed some years of familiarity with energy before being called upon to grapple with entropy. In some approaches, the meaning of entropy is illustrated by an appeal to statistical mechanics where it is possible to relate entropy to the number of microscopic configurations of the system which are consistent with a single macroscopic (thermodynamic) state. In keeping with the declared purpose of this book, we shall not base our arguments on discussions of this kind.

Nevertheless, it is important to take some steps to make the bare statement of the second law contained in Equations (4.1) and (4.2) more satisfying intuitively. Such steps can in no way lead to a proof of the law but may make it seem more plausible. Since we need not concern ourselves with a rigorous proof, many liberties may be taken with the plausibility argument. Any path which maximizes our intuitive satisfaction is permissible. The next few sections will be devoted to pursuit of such a path.

2. Need for an Additional Extensive Function of State

A useful method for describing the concept of entropy is to show how it arises in a natural manner almost out of necessity.

Consider Figure 1.2, modified to the extent that the rigid walls of the container are made adiabatic rather than diathermic so that the system in the container is no longer in contact with the thermostat. Then the system is *isolated* in the thermodynamic sense. Neither energy nor matter may be exchanged between the system and its surroundings. *In spite of this, one may by external intervention induce a change in the system.* This may be performed by opening the lock, an act which may be accomplished without exchanging energy with the system since the lock and walls are not part of the system. We are left in the peculiar position of being able to influence the behavior of the system without being able to transmit to it a signal consisting of energy or matter or both!

In the example cited, the opening of the lock represents the removal of a constraint (see Chapter 1, Section 8); and the ability to influence a system without exchanging energy or matter is a characteristic feature of a constraint. In Figure 1.2 the process induced by opening the lock will involve the expansion of the gas on the left of the partition and the movement of the partition to the right. If the partition is frictionless and the gas itself exhibits no internal friction, then the partition will simply oscillate without being damped. The situation is similar to that which would occur in connection with the pendulum of Figure 1.3 if it were equipped with frictionless bearings. It is possible to remove the constraining force $\mathcal{X}_1$ without exchanging energy with the pendulum, so that it oscillates without being damped. It is even possible to reapply the force $\mathcal{X}_1$ at a propitious moment so that the pendulum becomes confined to its original state. In this case, *the re-establishment of the constraint, like its removal, involves no exchange of energy with the pendulum.* The same is true of the frictionless oscillating gas described above. The lock may be closed and the partition confined to its original position without exchanging energy with the gas.

This would not be the case if either the gas or the pendulum possessed friction. In this case, it is a matter of experience that the constraints, once removed, cannot be reapplied without the expenditure of work. The frictionless systems may be referred to as *purely mechanical* systems whereas those with friction are *thermodynamic* systems.

This inability to re-establish a constraint, once it has been removed, without the expenditure of work is perhaps the most important feature which distinguishes a thermodynamic from a mechanical system. In fact, one might come very close to providing a satisfactory definition of thermodynamics by describing it as the study of the equilibrium properties of such systems.

Although we have used the modified (adiabatic) system of Figure 1.2 to illustrate how a change may be induced in an isolated system by the removal of a constraint, the situation is quite general; and one may think of many examples involving other systems and other constraints. One may inquire as to how we know that a change *will* take place when a constraint is lifted. The answer is somewhat circular. We assume that we can recognize a constraint for what it is. Then upon its removal (since thermodynamics is not concerned with rates of change), we are permitted an infinite period of time during which to wait for the completion of the induced process. We assume that given enough time, it will take place. If it does not, then by definition we have not removed the constraint.

Thus far, in the unfolding of this book, the concept of the constraint has come into focus again and again. It should already be clear that it plays a dominant role in thermodynamics, and it is natural to inquire into whether the concept can be quantified so that a precise measure of the degree of constraint of a system may be given. The mere existence of induced processes in isolated systems indicates the possibility of ordering its states according to degree of constraint. An induced process then carries the system from a more to a less constrained state; that is, the process is induced through the removal of a constraint.

We prefer to use the term "induced process" in place of the more conventional "spontaneous process." It is clear that spontaneous processes in thermodynamic systems must in fact be "induced" by the removal of a constraint. An induced process in an isolated system is obviously a real process and not a quasistatic one. Its path is not a sequence of equilibrium states plottable in state space. If the system "remains at equilibrium" during the change, then by definition no change is induced! As a result, Equation (4.2) must refer to induced processes. Since they are not reversible, they may be called *irreversible* processes.†

If an index can be found which measures the degree of constraint of a system, it will of course be a function of the state of the system. Furthermore, since the independent variables and functions of state may be used interchangeably (see Chapter 3, Section 1), the degree of constraint should also be capable of serving as a variable of state. Since the states of the system are ordered with regard to degree of constraint by referring to the occurrence of induced processes in isolated systems, the variable which measures the degree of constraint must correspond to constraints which may be applied to isolated systems.

A careful examination of the question reveals that constraints in *isolated* systems act only to fix extensive variables. For example, in the modified (adiabatic walls) system of Figure 1.2, it is the volume of the gas on the left of the piston which is fixed by the constraint, and volume is an extensive variable. In contrast, if we wanted to constrain the system to a given value

† The term *irreversible* also draws attention to the fact that in a thermodynamic system a constraint, once removed, cannot be reapplied (the deed cannot be undone or reversed) without the expenditure of work.

of temperature (an intensive variable), we would have to replace the adiabatic walls by diathermic ones so that the system could establish thermal contact with the thermostat. Under these conditions, the system is no longer isolated. The situation is quite general—intensive variables can be constrained only in nonisolated systems. For example, if one wanted to constrain the pressure, it would be necessary to have the system establish *mechanical* contact with a constant pressure ambient; and so an invariant pressure could not be established in an isolated system.

As a result of this situation, we are led to seek a new extensive function of state whose magnitude measures the degree of constraint. In anticipation of its discovery, let us denote the new function of state by the letter C, and name it the *degree of constraint.*

3. The Degree of Constraint

In the interest of pursuing the most convenient intuitive path, let us specialize our thinking to the simple system illustrated in Figure 4.1(*a*). This consists of a gas trapped in a cylinder closed by a piston. The piston furnishes one means of applying a constraint. Suppose at first that the cylinder and piston are so constituted as to isolate the enclosed system. If the piston is suddenly displaced outward at an infinite rate, the enclosed gas will expand (Figure 4.1(*b*)) without performing any work. As a result, the expansion will have occurred with no change of internal energy, provided that the walls of the cylinder and the material constituting the piston are adiabatic. When the piston comes to rest at a new position displaced to the right, it is clear that the degree of constraint will have been diminished. Not only has an induced process taken place, but the degree of confinement of the gas has been lessened. How can we measure in a meaningful way this reduction in degree of constraint?

One attractive possibility which presents itself immediately involves the measurement of the work which must be expended in order to re-establish the initial state of constraint. In the example under consideration, this work will have to be performed through a movement of the piston to the left so that it assumes its original position. Somehow we feel intuitively that the

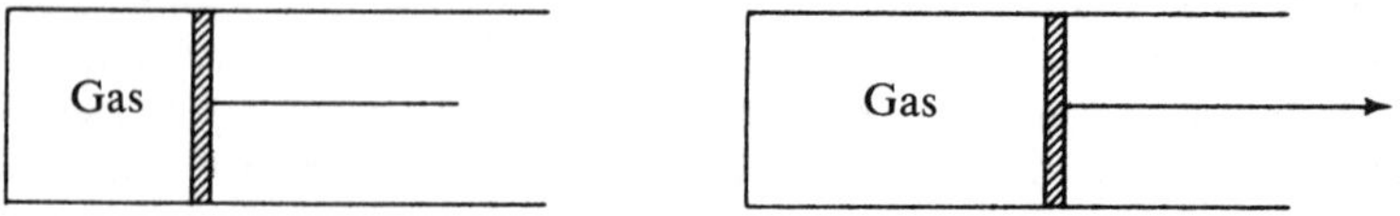

(a) (b)

FIGURE 4.1. *Illustration of the measurement of the degree of constraint.*

greater the work which must be expended in establishing a constraint the greater must be the degree of constraint. Although this is reasonable for a start, several improvements are possible.

Upon returning to the system of Figure 4.1 and considering the work necessary to restore the original constraint through a movement of the piston to the left, we see that the temperature of the gas presents a problem. As a matter of experience, we know that if the temperature is higher, more work will have to be expended in restoring the piston to its original position. This follows from the fact that the pressure of the gas against which the piston must work will be greater at the higher temperature. Since we wish the degree of constraint to be determined by extensive quantities which can in fact be constrained in an isolated system (in this case the volume determined by the position of the piston), we must somehow contrive to remove the influence of the intensive variables, temperature and pressure. One way of accomplishing this would be to divide the work performed in moving the piston to the left by some function of the empirical temperature θ defined in Equation (3.13). At the moment, no loss of generality is incurred by leaving this function of θ completely arbitrary and denoting it by $g(\theta)$.

By following the notation of Chapter 3, we denote the work performed on the system in moving the piston from right to left by $-Dw$. The form of $g(\theta)$ might then be chosen so that when the gas becomes hotter, $g(\theta)$ increases by just the amount required to maintain $-Dw$ divided by g constant for the desired increase of constraint. Thus, although $-Dw$ increases, so does $g(\theta)$, and the change of constraint dC which is now represented by

$$dC = -\frac{Dw}{g(\theta)} \tag{4.4}$$

remains independent of temperature. Since C is to be a variable of state, dC must be capable of being plotted in state space. This means that the path to which $-Dw$ applies must also be plottable in state space. For this reason we must restrict our attention to reversible processes, and $-Dw$ then refers to reversible work. This restriction in Equation (4.4) is also necessary because temperature can be defined operationally only in systems at equilibrium (at least in classical thermodynamics).

Substitution of Equation (3.18) into Equation (4.3) yields

$$dC = \frac{dU - Dq}{g(\theta)}. \tag{4.5}$$

Now the compression under consideration is the reverse of an expansion which might have occurred in an isolated system. That is, it is the reverse of a process in which the internal energy remains fixed, and therefore dU in

Equation (4.5) may be set equal to 0. Thus, we arrive at the result

$$dC = -\frac{Dq}{g(\theta)}, \tag{4.6}$$

in which Dq represents heat absorbed by the system during the reversible change. Incidentally, in the reverse process (compression) under consideration, the isolation of the system is relaxed because work (of compression) is performed on it. Furthermore, it will be necessary to replace the adiabatic walls with diathermic ones so that heat can be removed from the system as work is performed, and the internal energy may remain fixed. In the present case, the quantity $-Dq$ represents the heat removed.

4. Entropy

A method has thus been described for measuring the variable C which we call the *degree of constraint*. According to this method, the difference $C_2 - C_1$ between the degrees of constraint C_2 and C_1 corresponding to two thermodynamic states 1 and 2 is simply

$$C_2 - C_1 = -\int_1^2 \frac{Dq}{g(\theta)}, \tag{4.7}$$

where the integral is taken over some path in state space connecting states 1 and 2. Now the argument leading to Equation (4.7) has been developed through reference to the particular system of Figure 4.1. The constraint which has been considered (position of the piston) has also been a special one. Nevertheless, the integrand in Equation (4.7) is composed of quantities which have meaning in any system, and it is therefore possible to adopt (4.7) for the measurement (at least formally) of the degree of constraint in any system. This generalization is made more reasonable when it is considered that the degree of constraint as it has been defined is related to the work which must be expended (in any system) to apply the constraint.

It may be possible to induce a series of sequential changes in an isolated thermodynamic system by lifting one constraint after another so that C continually diminishes. Since constraints restrict the behaviors of systems, they tend to confine and organize them. From this viewpoint, the system undergoing a sequence of changes induced by the continuous lifting of constraints becomes increasingly disorganized. Since these changes occur within isolated systems, one may conclude that varying degrees of organization may correspond to the same total energy, and that induced changes at constant energy always take place in the direction of greater disorganization. It follows that an *increase* in C corresponds to an *increase* in the state of organization of the system.

Since induced processes are made to occur in isolated systems through the *lessening* of a constraint, it now follows that one such process in an isolated system can cause a change from state 1 to 2 provided that

$$C_1 > C_2$$

or (4.8)

$$\Delta C < 0.$$

The discussion thus far has left us in the following position:

> For any thermodynamic system there exists a function of state which we shall call the *degree of constraint* and denote by the symbol C. The method for measuring the degree of constraint is based upon the fact that corresponding to any change of state for the system conducted *reversibly*
>
> $$dC = -\frac{Dq}{g(\theta)}, \tag{4.9}$$
>
> where Dq is the heat absorbed by the system during the reversible change and $g(\theta)$ is a definite but as yet unspecified function of the empirical temperature θ alone. Furthermore, for any change which can in fact take place in an *isolated* system,
>
> $$\Delta C < 0 \tag{4.10}$$
>
> (where Δ stands for a finite change).

Comparison of this statement with the statement of the second law, which appears at the beginning of this chapter, will reveal that the two statements are identical if the definition

$$S = -C \tag{4.11}$$

is introduced, and if the term *degree of constraint* is replaced in the latter statement by the word *entropy*. Thus, the rather abstract formulation of the second law which has been presented may now be invested with some physical meaning. For some purposes $-S$ may be more satisfying physically than S (certainly it is easier to think of the degree of constraint rather than the degree of lack of constraint), but the positive sign has been bequeathed by history and this is too late a time for change.

It can be shown that C is an extensive quantity through its relation to $-Dw$, the work performed on the system, in much the same way that the internal energy is shown to be an extensive quantity in Equation (3.27) through its relation to the work performed. Since S is merely the negative of C, it follows that S is also an extensive quantity.

The determination of U is facilitated by the measurement of work performed in an adiabatic process, whereas according to Equation (4.9), the determination of S requires the measurement of the heat absorbed by the system in a reversible process. Although the *work* performed during the *adiabatic* process is usually measured directly, the same is not true for the *heat* absorbed during a *reversible* process for the simple reason that a reversible process is one which cannot in fact occur except in a limiting manner when the process is conducted very slowly. This point was discussed in Chapter 1, Section 7 in connection with the discussion of reversible or quasistatic processes. It was noted that the *work* performed during the reversible process could in any event be computed if the equation of state for the system was known. A similar situation holds for the *heat* absorbed in a reversible process; it can be computed if enough is known about its state behavior.

For example, suppose the system is a simple fluid. Once the mass is fixed the variables of state will be two in number. These can be selected conveniently as the volume V and temperature θ. This system can only perform mechanical work on its environment through a change in its volume. If the volume changes by the small amount dV and the pressure is p, the work done by the system will be

$$Dw = p\,dV. \tag{4.12}$$

Substitution of Equation (4.12) into Equation (3.18) yields, for the heat absorbed during the reversible change,

$$Dq = dU + p\,dV. \tag{4.13}$$

Since dU is an exact differential, it can be expressed in terms of the independent variables θ and V. Thus,

$$dU = \left[\frac{\partial U}{\partial \theta}\right]_V d\theta + \left[\frac{\partial U}{\partial V}\right]_\theta dV. \tag{4.14}$$

By combining this with Equation (4.12) we get

$$Dq = \left[\frac{\partial U}{\partial \theta}\right]_V d\theta + \left(p + \left[\frac{\partial U}{\partial V}\right]_\theta\right) dV, \tag{4.15}$$

from which we see that Dq is in the form of a Pfaff differential expression of the sort illustrated in Equation (2.2). Since dS is an exact differential (by postulate, since S is supposed to be a function of state), the second law amounts to the statement that $g(\theta)$ is an integrating denominator for the Pfaff differential Dq. We shall have more to say about this when we consider the principle of Carathéodory. In any event, it should now be clear from Equation (4.15) that the heat Dq, absorbed reversibly, can be expressed in terms of functions of state or their derivatives. If the state behavior for the system is known, p, $\left[\frac{\partial U}{\partial V}\right]_\theta$, and $\left[\frac{\partial U}{\partial \theta}\right]_V$ are also known so that Dq is known even though it refers to a reversible, and therefore hypothetical, process. Thus, S can be determined to within a constant.

5. Extremal Properties of the Entropy

If we remove a series of constraints and process after process is induced in the isolated system, the entropy progressively increases. Stated another way, if we pass to states of equilibrium which are determined by fewer and fewer variables, the entropy progressively increases. If at some stage we refuse to remove any more constraints, then the entropy of the isolated system is maximized subject to the constraints we insist upon retaining, that is, for an equilibrium controlled by a certain minimum number of variables. Whatever we do to this system, if we do not lessen the degree of constraint we cannot increase its entropy as long as it remains isolated.

As we have indicated in Section 4, to perform an *inverse* process we must relax the isolation and contrive to make the internal energy, before and after the inverse process, the same. Otherwise, the process cannot be the inverse of one which occurs in an isolated system, for which the energy of the initial and final states must be the same.

If an isolated system has come to equilibrium subject to certain *retained* constraints, then for any process which is the inverse of one which could have been involved in reaching this equilibrium

$$\Delta S < 0. \tag{4.16}$$

Sometimes one sees the statement:

> If for all possible changes in an isolated system $\Delta S < 0$, then the system must be in equilibrium.

This is loose and misleading. It is meant to correspond to the statement we have just made in connection with Equation (4.16). But as we have seen, the change to which Equation (4.16) corresponds cannot in fact occur in an *isolated* system. It can correspond to the *inverse* of one which takes place in an isolated system. To realize the process, the isolation must be destroyed.

The fact that the entropy is maximized when equilibrium is established subject to certain constraints, and that, as a consequence Equation (4.16) holds, means that for every suitable infinitesimal *inverse* process the differential of the entropy may be set equal to zero. This is nothing more than the usual mathematical condition for the existence of an extremum (Chapter 2, Section 6). As a result, the correct statement should be:

> If a system is in equilibrium subject to certain constraints, then for every infinitesimal variation during which the original constraints are retained and which is the *inverse* of a process which could have occurred under conditions of isolation, $\delta S = 0$.

6. Virtual Variations

The variation referred to in the last paragraph of the previous section is termed a *virtual* variation. Differentials which correspond to virtual variations employ the symbol δ rather than d. The term "virtual variation" has occasioned some confusion in thermodynamics, and it is worthwhile to spend a moment discussing it.

The most serious misimpression (and one which has been propagated by some very distinguished authors) is the idea that the virtual variation is so general that it may carry the system from a state of equilibrium to one of *nonequilibrium*. Nothing could be further from the truth. For example, a thermodynamic property like entropy, a function of state, could not be defined in nonequilibrium (at least not by the methods of classical thermodynamics) and so the variation δS could not be defined, since it would have to measure the difference between the entropy in the final (nonequilibrium) state and the initial one. Then what does the virtual variation refer to?

The answer is quite simple. In Section 8 of Chapter 1, we indicated that every state of equilibrium is established subject to certain constraints, and that these constraints bear a one-to-one correspondence to the independent variables of the system in the state of equilibrium thus established. These constraints were symbolized by χ_0. It was also shown that the system could be displaced to a new state of equilibrium through the introduction of an additional constraint (or constraints) which were symbolized by χ_1. During the displacement the original constraints χ_0 are retained, and the new state is therefore established subject to the combined set involving both χ_1 and χ_0. However, this new state, having been established subject to a larger number of constraints, possesses more independent variables than the old one. In Chapter 1 the situation was illustrated in Figure 1.3 by appeal to an example drawn from mechanics.

The *virtual* variation to which reference has been made above is of precisely this sort. For example, when the entropy is maximized subject to those constraints which we insist upon retaining, it is maximized subject to a set χ_0. The virtual variation is then achieved through the introduction of

$\mathcal{X}_1$ ($\mathcal{X}_0$ being retained) so that the system is displaced to a new *equilibrium* state. This new state, however, is characterized by more variables than the original one.

In the particular example of an *isolated* system, among the set $\mathcal{X}_0$ are the conditions that U be constant and that V be constant, so that the system performs no volume work.† Consider the system of Figure 1.2, modified to have adiabatic walls so that it may be isolated. In this case, the total volume of the system $V = V_1 + V_2$ must remain constant. After the partition is unlocked, equilibrium is established subject to constant U and V (and also constant mass). A virtual variation from this equilibrium—for example, returning the partition to its original position—is carried out at constant U and V (although the constancy of U will have to be maintained, as explained in connection with Figure 4.1, by introducing diathermic walls) through the introduction of another constraint $\mathcal{X}_1$ which really corresponds to the replacement of the lock. Thus, the set $\mathcal{X}_0$ is retained and the system is augmented.

The important thing to note is that the variation *does not* carry the system from equilibrium to nonequilibrium, but rather from an equilibrium state characterized by a given number of variables to another *equilibrium* state controlled by a larger number of variables.

The term "virtual" is used to emphasize the fact that in the application of an equation such as $\delta S = 0$, the detailed character of the more highly constrained state to which the system is displaced is never invoked. The equation $\delta S = 0$ is a differential equation which may be represented in terms of the differentials of the independent variables. As we shall see, it is possible to use the condition $\delta S = 0$ to establish relations among the coefficients of these differentials. These relations provide interconnections among the macroscopic properties of the system. But the coefficients of the differentials refer only to the *initial* state of the variation, *not* to the *final* one. Hence, the properties of the final state are not invoked, except insofar as their existence gives meaning to the variation.

Since the precise nature of the set of constraints $\mathcal{X}_1$ is not invoked, $\mathcal{X}_1$ may be very general indeed. In fact, one may even employ constraints which cannot be applied, in practice, in any reasonable experiment. It is this inability at times to apply the constraints in anything but a thought experiment which gives rise to the term "virtual." In spite of this, the displacement is intended to carry the system to a new *equilibrium* state, even though the state may exist only in abstraction.

† It is not necessary to require that the system be enclosed in adiabatic walls because the combination of constant U and no exchange of work implies this by virtue of the first law. Notice that the requirement of U being constant does not by itself demand that the system be isolated.

There are situations in thermodynamics in which the real properties of the final state are required. These do not arise, however, in connection with the direct use of extremal conditions like $\delta S = 0$.

7. Temperature Scale and Thermodynamic Efficiency

As a practical matter, temperature scales are usually established with reference to some measurable property of a real substance. The substance itself can then be used as a thermometer in the manner suggested by the discussion of Chapter 3, Section 2. In Chapter 5 we shall devote some time to the discussion of real (or at least idealized real) systems, taking special care to distinguish thermodynamic from nonthermodynamic information. In that chapter, the role of the *ideal* gas as a thermometer will be elaborated. For the present, it is convenient to advance our discussion, when possible, without reference to a particular substance. In this connection, the second law makes possible the establishment of a temperature scale and the identification of $g(\theta)$ without such reference. This scale is known as the "thermodynamic scale of temperature" and, as we shall see later, is closely related to the ideal gas scale. We start with the arbitrary function $g(\theta)$ of the empirical temperature θ. The following discussion is facilitated by reference to Figure 4.2.

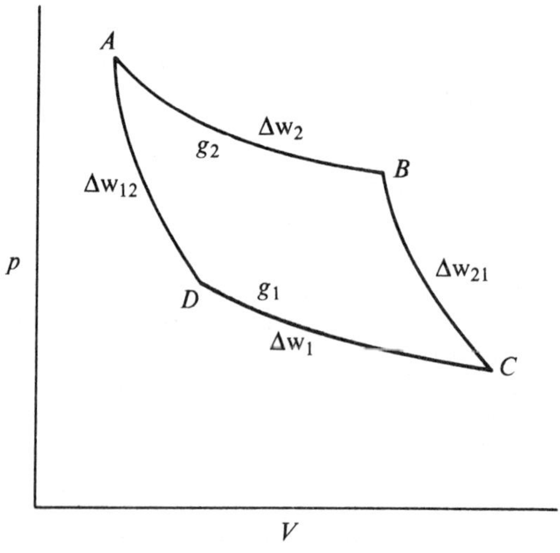

FIGURE 4.2. *Diagram of typical Carnot cycle.*

Consider *any* thermodynamic system at equilibrium and in contact with a thermostat at some temperature g_2. Call this state of the system state A. Let the system undergo a *reversible* change to state B (while still in contact with the reservoir at temperature g_2) such that it performs work Δw_2 on its environment. For example, if the system were the gas enclosed in the cylinder of Figure 1.1, the change from A to B might be the reversibly expansion discussed in connection with that figure. Alternatively the system might be an electrochemical cell, and the change might be a process of discharge of the cell during which it performs electrical work on its environment. The precise specification of the system is not important here. It must only be possible to set it up so that it exchanges heat and work with its surroundings in the manner demanded.

The change from state A to state B is *isothermal*; the system remains in contact with the thermostat and its temperature is held constant at g_2. Not only does it perform the work Δw_2, but it absorbs the heat Δq_2 from the environment.

From B the system is carried *reversibly* to state C by causing it to undergo an *adiabatic* change. To accomplish this the system will have to be isolated from the thermostat by being enclosed in adiabatic walls. Let the adiabatic change be such that the system is cooled in passing from state B to C so that its temperature at C is g_1. Of course, no heat can be exchanged with the surroundings in the passage from B to C, but the system may perform the work Δw_{21}.

Next, place the system in contact with a thermostat at temperature g_1 and cause it to undergo another *isothermal* reversible change, this time from state C to D. The work performed on the environment will be Δw_1 and the heat absorbed will be Δq_1. Finally, the system is removed from contact with the thermostat at g_1 and returned *adiabatically* and *reversibly* to the original state, thus completing the cycle. During this last change (from D to A), the work performed may be denoted by Δw_{12}.

The reversible process just described is one example of a Carnot cycle and is diagrammed schematically in Figure 4.2 for the case where the working substance is a simple gas with variables p and V. The Carnot cycle employing an ideal gas as the working substance has often been used to suggest the existence of the entropy function. By utilizing the *special* properties of the ideal gas (see Chapter 5), it is possible to show that the sum of the various heats absorbed divided by the respective temperatures at which they are absorbed is zero for the complete cycle. The cycle leaves the system in its original state, and this suggests that the sum has been a sum of changes of a function of state, whose increments can be measured by the ratios of the heats absorbed reversibly, to the temperatures. This function, of course, is assumed to be the entropy. However, as we shall show in Section 11 for a two-variable system, the existence of the entropy is a direct consequence of the *first law*. To demonstrate something new, one must employ at least a three-variable system.

Upon returning to the cycle described above, we see that the first law (Equation (3.18)) demands, for the entire cycle,

$$\begin{aligned} U &= \Delta q_2 + \Delta q_1 - \Delta w_2 - \Delta w_{21} - \Delta w_1 - \Delta w_{12} \\ &= \Delta q_2 + \Delta q_1 - \Delta w = 0, \end{aligned} \tag{4.17}$$

since no net change in internal energy can be realized as the system returns to its initial state. Here we have set

$$\Delta w = \Delta w_2 + \Delta w_{21} + \Delta w_1 + \Delta w_{12} \tag{4.18}$$

to symbolize the total work performed over the cycle. Since the cycle has been performed reversibly, the second law (Equation (4.1)) requires

$$\frac{\Delta q_2}{g_2} = \Delta S_2, \qquad \frac{\Delta q_1}{g_1} = \Delta S_1, \tag{4.19}$$

where ΔS_2 and ΔS_1 are changes in entropy. Since no net change in the entropy is possible when the system returns to its initial state,

$$\Delta S = \Delta S_2 + \Delta S_1 = \frac{\Delta q_2}{g_2} + \frac{\Delta q_1}{g_1} = 0. \tag{4.20}$$

The quantity,

$$\eta = \frac{\Delta w}{\Delta q_2}, \tag{4.21}$$

is usually called the efficiency of the cycle. The system carried around the cycle has been used as a machine. As we shall see, experiment demands that both g_2 and g_1 be of the same sign. Then Equation (4.20) requires Δq_1 to be negative if Δq_2 is positive. Thus, only Δq_2 represents heat absorbed *from* the environment; Δq_1 is actually heat returned *to* the environment. With this requirement, the meaning of η becomes clear. It represents the *fraction* of the heat *absorbed from* the environment which is converted into work by the machine operating reversibly in cycles between the two thermostats at g_2 and g_1, respectively. The heat, which is not converted to work, is returned to the surroundings as Δq_1 through the medium of the thermostat at g_1.

By substituting, for Δw in Equation (4.21), the value $\Delta q_2 + \Delta q_1$ offered by Equation (4.17) yields

$$\eta = \frac{\Delta q_2 + \Delta q_1}{\Delta q_2} = 1 + \frac{\Delta q_1}{\Delta q_2}. \tag{4.22}$$

According to Equation (4.20),

$$\frac{\Delta q_1}{\Delta q_2} = -\frac{g_1}{g_2}. \tag{4.23}$$

Substitution of this into Equation (4.22) gives

$$\eta = \frac{g_2 - g_1}{g_2}. \tag{4.24}$$

Now η has a perfectly definite physical meaning. Furthermore, the relation (4.24) is universal, since it applies to any working substance carried reversibly around the cycle. Equation (4.24) may be rearranged to display the functional dependence of g_1 upon g_2. Thus,

$$\frac{g_1}{g_2} = (1 - \eta). \tag{4.25}$$

From Equation (4.25) it is clear that *the efficiency* η *determines the ratio of the temperatures of the two thermostats.* This ratio is independent of the working substance because η is independent, nothing having been specified about the substance, in our arguments. Thus, the arbitrary fixing of *one* point on the scale, say the value of g_1, determines the values of all other temperatures g. All that is necessary is the measurement of the efficiency η of a Carnot cycle, operating between thermostats at temperatures g and g_1 and the application of Equation (4.25), with g_2 replaced by g. In this way, g is related to a perfectly definite measurable physical quantity. Thus, a temperature scale can be established having no relation to a particular substance. This scale is known as the thermodynamic scale of temperature. If g_1 corresponds to a thermostat consisting of melting ice under its own vapor pressure and is set equal to 273.16, we thereby define the ideal gas or Kelvin temperature T. We shall demonstrate this in Chapter 5.

8. Maximum Efficiency

The Carnot cycle was developed originally for the purpose of investigating the efficiencies of machines operating in cycles. The concept of entropy was a natural outgrowth of the various considerations surrounding this cycle. One result of these early deliberations was the realization that the efficiency η specified by Equation (4.24) and applicable to a reversible cycle is the maximum which can be achieved with a machine operating between heat reservoirs at temperatures g_2 and g_1, respectively. This fact can be assumed as a postulate (in fact as a part of the second law), or it can be derived from the second law in the form we have chosen to adopt.

To prove this, assume that every part of the *environment* which contacts the machine utilized in a Carnot cycle behaves reversibly no matter how fast the actual cycling is performed. No loss of generality is incurred by this device because the state of a thermodynamic system, once defined, is determined only by its own variables. One might ask, however, whether it is possible in fact to provide an environment which always behaves reversibly. The answer is a relative one. If the environment consists of systems which come to equilibrium very rapidly compared to the rate at which processes take place in the cycled machine, then in an approximation which depends upon how great these differences of rate are, it is possible for the environment to behave reversibly.

For example, suppose two compartments containing gases at different pressures are separated by a plug consisting of some viscous fluid capable of generating a great deal of internal friction. As a result of the pressure differential the plug will deform, and during this motion there will be a non-equilibrium stress distribution in the fluid and an irreversible dissipation of energy by means of internal friction. Therefore, the flow of the plug cannot be described as a sequence of equilibrium states (the irreversible path cannot be plotted in state space). On the other hand, the gases relax into equilibrium so rapidly and with so little internal friction compared to the rate of deformation of the plug that their pressures are always uniform, and the respective processes of expansion and compression may be regarded as reversible. Thus, two systems may be in mechanical and thermal contact, and one may behave reversibly while the other does not. It is a matter of relative rates.

Upon returning to our machine, and assuming its environment always to behave reversibly, we may continue the argument. It will be convenient henceforth to use the symbol T for the temperature in place of g in anticipation of the choice of scale factor in Equation (4.24). Denote by η_{irr}, the efficiency for a cycle conducted irreversibly, reserving η for reversible processes. Thus,

$$\eta_{\mathrm{irr}} = \frac{(\Delta w)_{\mathrm{irr}}}{(\Delta q_2)_{\mathrm{irr}}}, \tag{4.26}$$

where the abbreviation irr is appended to both the work and heat to indicate that these correspond to irreversible processes. We wish to prove that for a machine operating in cycles between reservoirs (in the environment) at temperature T_2 and T_1, respectively,

$$\eta_{\mathrm{irr}} < \eta, \tag{4.27}$$

or that the reversible cycle is the most efficient.

Now for the entire process, whether reversible or irreversible,

$$\Delta U = \Delta w - \Delta q = 0, \tag{4.28}$$

since U is unchanged over one cycle. By substituting Equation (4.28) into Equation (4.26) we obtain

$$\eta_{irr} = \frac{(\Delta w)_{irr}}{(\Delta q_2)_{irr}} = 1 + \frac{(\Delta q_1)_{irr}}{(\Delta q_2)_{irr}} = 1 + \alpha_{irr}, \tag{4.29}$$

where we have utilized the fact that since no heat is absorbed by the system during the adiabatic legs of the cycle, only the heats absorbed at T_1 and T_2 contribute to Δq,

$$\Delta q = \Delta q_1 + \Delta q_2. \tag{4.30}$$

From Equation (4.29)

$$\alpha_{irr} = \frac{(\Delta q_1)_{irr}}{(\Delta q_2)_{irr}}. \tag{4.31}$$

By the same line of argument for the cycle conducted reversibly,

$$\eta = 1 + \alpha, \tag{4.32}$$

where

$$\alpha = \frac{\Delta q_1}{\Delta q_2}, \tag{4.33}$$

where the heats now refer to reversible processes.

Let us make the further assumption that the system with its environment constitutes still a larger system which is isolated in the thermodynamic sense. It is clear that we can always accomplish this through the use of suitable rigid adiabatic walls.

Upon comparing Equations (4.29) and (4.32) with Equation (4.27), it is apparent that

$$\alpha_{irr} < \alpha \tag{4.34}$$

is equivalent to Equation (4.27). Now the change in entropy for the *environment* can be computed even when the machine behaves irreversibly, for the environment has been chosen to always behave reversibly. Thus, the heats absorbed *reversibly* by the environment are $-(\Delta q_1)_{irr}$ and $-(\Delta q_2)_{irr}$, since heat absorbed by the machine comes from the environment. According to Equation (4.1), therefore, the change in entropy for the environment is

$$\Delta S = -\frac{(\Delta q_1)_{irr}}{T_1} - \frac{(\Delta q_2)_{irr}}{T_2}. \tag{4.35}$$

Since the machine has covered a cycle, its entropy change is zero, independent of whether it behaves reversibly or not. Thus, for the combined system, machine and environment, the total change in entropy is prescribed by Equation (4.35). But the combined system by definition is isolated and has experienced an irreversible change since part of the system, the machine, has behaved irreversibly. Therefore, according to the second law, and especially to Equation (4.2),

$$\Delta S = -\frac{(\Delta q_1)_{\text{irr}}}{T_1} - \frac{(\Delta q_2)_{\text{irr}}}{T_2} > 0$$

or

$$\frac{(\Delta q_1)_{\text{irr}}}{(\Delta q_2)_{\text{irr}}} = \alpha_{\text{irr}} < -\frac{T_1}{T_2}. \tag{4.36}$$

On the other hand, for the cycle conducted reversibly we have from Equation (4.20), with T substituted for g

$$\frac{\Delta q_1}{\Delta q_2} = \alpha = -\frac{T_1}{T_2}. \tag{4.37}$$

Substitution of α from Equation (4.37) for $-(T_1/T_2)$ in (Equation 4.36) yields $\alpha_{\text{irr}} < \alpha$; the same as relation (4.34), equivalent to (4.27) which we set out to prove originally.

Thus, maximum efficiency is achieved when the cycle is conducted reversibly.

9. Additional Use of the Reversible Environment

The device of viewing the system and its environment as subsystems of a single *isolated* system in which the environment subsystem behaves reversibly finds additional use in thermodynamic argument. Consider the system as it undergoes some infinitesimal change, not necessarily cyclic. If the change is made to occur reversibly, then the heat absorbed by the system is specified by the second law (see Equation (4.1)) to be

$$Dq = T\,dS, \tag{4.38}$$

where dS represents the change in entropy for the system. If the change occurs irreversibly, we denote the heat absorbed by the system by Dq_{irr}.

Now the heat absorbed by the environment during such an irreversible change is

$$Dq_e = -Dq_{\text{irr}}, \tag{4.39}$$

where the subscript e refers to "environment." Since all changes take place reversibly in the environment, the change of entropy dS_e can always be computed independent of whether or not the system behaves reversibly. Thus, according to Equation (4.1),

$$dS_e = \frac{Dq_e}{T} = -\frac{Dq_{\text{irr}}}{T}. \tag{4.40}$$

The total change in entropy for the isolated system is

$$dS + dS_e = dS - \frac{Dq_{\text{irr}}}{T} > 0, \tag{4.41}$$

where the inequality (4.2) has been utilized. The quantities to the right of the equals sign refer only to the system and not to its environment (the environment has been utilized only for the purpose of establishing Equation (4.40)); and the relation is therefore valid, independent of the nature of the environment. Multiplication by the positive quantity T yields

$$T\,dS - Dq_{\text{irr}} > 0. \tag{4.42}$$

Since $T\,dS$ (according to Equation (4.38)) represents the heat absorbed when the same change is conducted reversibly, it becomes apparent from Equation (4.42) that for a given change the system absorbs the maximum heat when the change is conducted reversibly.

A similar conclusion may be arrived at in connection with the work Dw_{irr} performed by the system. From the first law (3.18) we have

$$Dq_{\text{irr}} = dU + Dw_{\text{irr}} \tag{4.43}$$

and

$$Dq = T\,dS = dU + Dw, \tag{4.44}$$

where Equation (4.44) refers to the same change conducted reversibly; and dU, being an exact differential, is the same in Equations (4.43) and (4.44); that is, it is independent of whether the change is conducted reversibly or not. Substitution of Equations (4.43) and (4.44) into Equation (4.42) yields

$$Dw - Dw_{\text{irr}} > 0. \tag{4.45}$$

Thus, the maximum amount of work is performed *by* the system, in a given change, when that change occurs reversibly.

10. Conversion of Heat into Work

Equation (4.24) reveals that a machine operating in cycles cannot achieve unit efficiency (even when it performs reversibly) unless the lower temperature of the cycle g_1 (or T_1) is zero. However, according to Equation (4.38), no heat may be exchanged with a reservoir with $T = 0$, so that a cycle with $T_1 = 0$ cannot be completed. Therefore, for a machine operating in cycles, perfect efficiency cannot be achieved. Hence, under no realizable conditions can a machine operating in cycles convert heat entirely into work.

On the other hand, this restriction is inapplicable to noncyclic processes. Thus, suppose a substance existed whose internal energy U depended only upon its temperature T.† If it is a simple substance like a gas, it may perform

† In Chapter 5, we shall show that an ideal gas possesses this property.

work on its environment through expansion. In fact, the gas may be enclosed in the cylinder of Figure 1.1 and may be subjected to an isothermal expansion. Since its temperature remains fixed, $\Delta U = 0$ and the first law, Equation (3.18), demands

$$\Delta w = \Delta q, \tag{4.46}$$

so that in this process heat is converted entirely into work.

After the process has taken place, however, the piston in Figure 1.1 has moved to a higher level and the gas is at a lower pressure. In contrast, a machine operating in cycles will return to its original condition after each cycle, and some heat will have been converted into work. From the practical point of view, this is a distinct advantage because it is clear that in a system, such as that of Figure 1.1, the continued conversion of heat into work will result eventually in an enormous expansion and a severe reduction in gas pressure. Thus, even though heat has been converted entirely into work, the pressure will eventually become so low that it will be impractical to withdraw work from the system. Cyclic operation between two reservoirs at different temperatures overcomes this difficulty.

Thus, it is possible to convert heat entirely into work, but only by incurring the penalty that the system is not left in its original state after the process has been completed.

Of the several forms into which the second law has been cast, a traditional one is the so-called "principle of Thompson" which states that:

> There is no device which can transform heat withdrawn from a reservoir completely into work without at the same time undergoing other changes.

This will be recognized as a summary of the discussion of the last few paragraphs. From this statement, the existence and properties of the entropy function can be inferred.

Another traditional form, the so-called "principle of Clausius," states:

> There is no device which can transfer heat from a colder to a warmer reservoir in such a way that neither work is performed nor changes otherwise introduced in the components of the device.

From this it is also possible to infer the existence and properties of the entropy. However, both of these forms, as well as others which have been used, do not immediately expose the formal essence of the second law which is the assertion of the existence of a new function of state and a prescription for its measurement. It is for this reason that we have adopted the form which appears at the beginning of this chapter. This form contains such information explicitly. The two statements presented above can be derived from it.

11. The Principle of Carathéodory†

The arguments to which we have appealed, in this chapter, have placed systems in contact with environments which have contained certain idealized components such as infinite reservoirs, frictionless pistons, and surroundings which always behave reversibly. No loss of generality was incurred as a result of the employment of these components because they were confined to the environment, and the state of the system was determined by its own variables. Nevertheless, to the mathematician Carathéodory this situation was objectionable.

Thermodynamics was distinguished from all other scientific disciplines through its apparent dependence upon these idealizations. Furthermore, any statement of the second law which achieved immediate contact with physical intuition suffered, as do the two statements in the previous section, from the need to mention devices and machines. On the other hand, any statement cast in more formal mathematical terms, such as the one we have adopted, does not make immediate contact with physical intuition. Carathéodory sought to remedy this situation by supplying a statement which did not mention mechanical devices, and which was pleasing to the intuition. This form of the second law is known as "Carathéodory's principle."(5)

Carathéodory based his idea on the single assumption, derived from experience, that:

> In an arbitrary neighborhood of every point in macroscopic state space, there are adjacent points which cannot be reached from the first point by adiabatic means.

As an example, consider Figure 3.1. The shaded region marked "inaccessible" is inaccessible from P_0 by adiabatic means. If it were required that the means be reversible as well as adiabatic (so that the path of the change might be plotted in state space), the accessible region would be reduced from the unshaded *area* to a *curve* passing through P_0.

To see this, set $Dq = 0$ as would be required for an adiabatic process. With this limitation, Equation (4.15) can be rearranged to

$$\left[\frac{\partial V}{\partial \theta}\right]_{\text{adiabatic}} = \left\{\frac{\left[\dfrac{\partial U}{\partial \theta}\right]_V}{p + \left[\dfrac{\partial U}{\partial V}\right]_\theta}\right\}. \tag{4.47}$$

† In the author's view, Carathéodory's development is an elegant mathematical contribution but adds little to one's basic understanding of thermodynamics. In this sense, a discussion of Carathéodory is somewhat misplaced in a book of this kind, and we offer it only because there seems to be so much interest in the subject. The disinterested reader may omit it entirely and skip to Section 12 without loss of continuity.

The derivative on the left represents the slope of a reversible path in state space. The quantities $\left[\frac{\partial U}{\partial \theta}\right]_V$, $\left[\frac{\partial U}{\partial V}\right]_\theta$, and p in the expression on the right are uniquely defined at P_0 for the two-variable system involved in Figure 3.1. Hence, the derivative $\left[\frac{\partial V}{\partial \theta}\right]_{\text{adiabatic}}$ is also uniquely defined and generates a unique curve through P_0. Thus, for a reversible process the accessible points lie on a *curve* rather than in an area. It is clear that the reversible case corresponds to a specialization of the principle of Carathéodory. The domain of accessible points is even more limited than in the irreversible case.

Equation (4.15) is a Pfaff differential expression in two variables. It constitutes an example of Equation (2.2) limited to two variables (V and θ) with $D\psi$ replaced by Dq. According to the discussion in Section 2, Chapter 2, Pfaff expressions in *two* variables always possess integrating denominators. Thus, there exists a function of state, which we may denote by S, whose differential dS is defined by

$$dS = \frac{Dq}{T}. \tag{4.48}$$

It may be shown, without invoking the principle of Carathéodory, that T is identifiable with temperature, and S is the entropy as we have defined it. As a result, the *existence* of the entropy may be inferred for the two-variable system without appealing to the principle of Carathéodory or to any other form of the second law. As it stands then, for the two-variable system, the existence of entropy is a direct consequence of the first law since Equation (4.15) is a direct consequence. The crucial step depends upon the fact that a Pfaff expression in two variables always possesses an integrating denominator.

If we pass to a system involving three or more variables, expressions like Equation (4.15) for Dq can be derived; but it is not immediately clear that they possess integrating denominators, and some new information is required. This is supplied by the principle of Carathéodory. Figure 3.1 has been advanced as an example of this principle.

The system to which the figure corresponds possesses only two variables. However, it is easy to conclude, through an examination of many real systems, that the same considerations of inaccessibility would apply even if three or more variables were involved. For example, the use of a mixture of two or more fluids in the adiabatic container would add variables of composition; but it is clear, intuitively, that similar regions of inaccessibility would exist.

The principle of Carathéodory specialized to the reversible case implies that corresponding to *any* point in state space there are other points, arbitrarily close, which are inaccessible from the first point along a solution curve of the Pfaff differential equation $Dq = 0$ (that is along an adiabatic path). But then the *theorem* of Carathéodory (Chapter 2, Section 2) demands that Dq possess an integrating denominator. It is necessary to show that the exact differential which therefore arises is the differential of entropy and that the integrating denominator is the thermodynamic temperature. Furthermore, the property embodied in the inequality (4.2) must be demonstrated. When these facts emerge as a consequence of Carathéodory's principle, it becomes evident that it is equivalent to the second law.

Consider two simple substances in thermal contact whose combined variables of state are V_1, V_2, and θ, where V_1 and V_2 are the respective volumes of the subsystems 1 and 2, and θ is the common empirical temperature. Thus, we deal with a system possessing three variables. Then for each of the subsystems, Equation (4.15) may be written with θ substituted for T. The heat term for the combined system is

$$Dq = Dq_1 + Dq_2 = \left\{\left[\frac{\partial U_1}{\partial V_1}\right]_\theta + p_1\right\} dV_1 + \left\{\left[\frac{\partial U_2}{\partial V_2}\right]_\theta + p_2\right\} dV_2 + \left\{\left[\frac{\partial U_1}{\partial \theta}\right]_{V_1} + \left[\frac{\partial U_2}{\partial \theta}\right]_{V_2}\right\} d\theta. \quad (4.49)$$

Carathéodory's principle, however, requires

$$Dq = Dq_1 + Dq_2 = \lambda\, d\varphi, \quad (4.50)$$

$$Dq_1 = \lambda_1\, d\varphi_1, \quad (4.51)$$

$$Dq_2 = \lambda_2\, d\varphi_2, \quad (4.52)$$

where the integrating denominators λ, λ_1, and λ_2, as well as the functions φ, φ_1, and φ_2, are functions of state. Combination of Equations (4.50), (4.51), and (4.52) yields

$$d\varphi = \left[\frac{\lambda_1}{\lambda}\right] d\varphi_1 + \left[\frac{\lambda_2}{\lambda}\right] d\varphi_2. \quad (4.53)$$

Instead of V_1, V_2, and θ, one may choose φ_1, φ_2, and θ as independent variables (φ_1 and φ_2 are functions of state). Then λ and φ appear as functions of φ_1, φ_2, and θ. Since $d\varphi$ is an exact differential, Equation (4.53) demands

$$\left[\frac{\partial \varphi}{\partial \varphi_1}\right]_{\varphi_2,\theta} = \frac{\lambda_1}{\lambda}, \quad \left[\frac{\partial \varphi}{\partial \varphi_2}\right]_{\varphi_1,\theta} = \frac{\lambda_2}{\lambda}, \quad \left[\frac{\partial \varphi}{\partial \theta}\right]_{\varphi_1,\varphi_2} = 0. \quad (4.54)$$

The last of these equations indicates that φ is independent of θ. Application of Equation (2.6) to Equation (4.53) yields

$$\left\{\frac{\partial}{\partial \theta}\left[\frac{\lambda_1}{\lambda}\right]\right\}_{\varphi_1,\varphi_2} = \left\{\frac{\partial}{\partial \varphi_1}\left[\frac{\partial \varphi}{\partial \theta}\right]_{\varphi_1,\varphi_2}\right\}_{\varphi_2,\theta} = 0,$$

$$\left\{\frac{\partial}{\partial \theta}\left[\frac{\lambda_2}{\lambda}\right]\right\}_{\varphi_1,\varphi_2} = \left\{\frac{\partial}{\partial \varphi_2}\left[\frac{\partial \varphi}{\partial \theta}\right]_{\varphi_1,\varphi_2}\right\}_{\varphi_1,\theta} = 0, \quad (4.55)$$

since $\left[\frac{\partial \varphi}{\partial \theta}\right]_{\varphi_1,\varphi_2} = 0$. Thus, $\frac{\lambda_1}{\lambda}$ and $\frac{\lambda_2}{\lambda}$ are also independent of θ.

Equations (4.55) may be rearranged as follows (where we omit the subscripts for notational simplicity):

$$\begin{aligned}\frac{\partial}{\partial \theta}\left[\frac{\lambda_1}{\lambda}\right] &= \frac{1}{\lambda}\frac{\partial \lambda_1}{\partial \theta} - \frac{\lambda_1}{\lambda^2}\frac{\partial \lambda}{\partial \theta} = 0,\\ \frac{\partial}{\partial \theta}\left[\frac{\lambda_2}{\lambda}\right] &= \frac{1}{\lambda}\frac{\partial \lambda_2}{\partial \theta} - \frac{\lambda_2}{\lambda^2}\frac{\partial \lambda}{\partial \theta} = 0,\end{aligned} \tag{4.56}$$

or

$$\frac{1}{\lambda_1}\frac{\partial \lambda_1}{\partial \theta} = \frac{1}{\lambda}\frac{\partial \lambda}{\partial \theta} = \frac{1}{\lambda_2}\frac{\partial \lambda_2}{\partial \theta}. \tag{4.57}$$

Now λ_1 is a function of state for the first substance and depends only on φ_1 and θ whereas λ_2 for the same reason depends only on φ_2 and θ. Equation (4.57) can therefore be true only if both $\frac{1}{\lambda_1}\frac{\partial \lambda_1}{\partial \theta}$ and $\frac{1}{\lambda_2}\frac{\partial \lambda_2}{\partial \theta}$ depend only upon θ. Otherwise, if Equation (4.57) were made to hold for a given set of values of φ_1, φ_2, and θ, by suitable adjustment of these variables, the equality could always be destroyed by the variation of φ_1 or φ_2 without the other. Thus, it follows from Equation (4.57) that

$$\frac{\partial \ln \lambda_1}{\partial \theta} = \frac{\partial \ln \lambda}{\partial \theta} = \frac{\partial \ln \lambda_2}{\partial \theta} = h(\theta), \tag{4.58}$$

where $h(\theta)$ is a universal function of θ, since it has the same value for two arbitrary substances, and equally so for the system which they constitute. Thus, $h(\theta)$ may be employed as a universal temperature function from which the customary temperature scale can be derived by a simple normalization of the integrating denominator.

If we now dispense with the indices and denote, by λ, the integrating denominator for an arbitrary system, it follows from Equation (4.58) that

$$\ln \lambda = \int h(\theta)\, d\theta + \ln \Phi, \tag{4.59}$$

where $\ln \Phi$ is a "constant" of integration still dependent upon the variables of state (other than θ), in this case upon φ. The clearing of logarithms yields

$$\lambda = \Phi(\varphi)e^{\int h(\theta)\, d\theta}, \tag{4.60}$$

so that for every thermodynamic system the integrating denominator can be decomposed into two factors, one of which depends upon temperature, while the other depends upon the remaining variables of state.

We now *define* the temperature by

$$T = Ke^{\int h(\theta)d\theta}, \tag{4.61}$$

where K is a constant.

Substitution of this definition into Equation (4.60) and the latter into Equation (4.50) yields

$$Dq = \lambda\, d\varphi = T\left[\frac{\Phi(\varphi)}{K}\right] d\varphi. \tag{4.62}$$

Therefore, we have shown that Dq can be factored into a temperature and the differential of some function of state. Thus, a new function of state, the entropy, may now be defined as

$$S = \frac{1}{K}\int \Phi(\varphi)\, d\varphi. \tag{4.63}$$

Through the proper choice of the normalization constant K, T may be made identical with the Kelvin temperature and S may be adjusted to the usual scale of entropy. With this definition Equation (4.62) becomes

$$Dq = T\, dS, \tag{4.64}$$

and the absolute temperature becomes the integrating denominator for the heat since dS is an exact differential.

Similar arguments may be advanced for more complicated systems in which Φ may depend upon additional variables beyond φ. It is not our intent to elaborate Carathéodory to this degree of detail but merely to indicate the method. Hence, the reader is referred to the original work for a a more comprehensive treatment.

On the other hand, we have still to derive the inequality (4.2) from Carathéodory's principle. We note that this inequality refers only to processes which can in fact occur in isolated systems, that is, to *induced* or *irreversible* processes. What we shall now show is that for any *adiabatic* process which is not reversible the entropy of a system must either always increase or always decrease. Since isolated systems are a subclass of adiabatic systems, the same result must apply to any change which can in fact occur in an isolated system. Since we only wish to indicate the method, the development may be continued using the composite system considered above, possessing the three variables V_1, V_2, and θ. Instead of θ, however, we may now employ S as the third variable.

Consider an adiabatic process in which the system proceeds from a state characterized by V_1^0, V_2^0, S^0 to a final state V_1, V_2, S. This process may be carried out in two stages: a reversible stage and an irreversible one. During the reversible stage the volumes are changed from V_1^0, V_2^0 to V_1, V_2. Since the process is adiabatic ($\Delta q = 0$), it follows from Equation (4.1) that no change of entropy is possible. Thus, the new state is characterized by V_1, V_2, and S^0. It should be pointed out that V_1 and V_2 can be varied independently and continuously. Thus, all combinations of V_1 and V_2 which leave the entropy unchanged may be reached by this reversible process.

If now we wish to change the entropy (maintaining the adiabaticity of the system), it is necessary to introduce an irreversible process—for a reversible process, $\Delta q = 0$ implies no change of entropy. This we do by fixing V_1 and V_2 and performing work by stirring, rubbing, et cetera, until the entropy has been changed from S^0 to S. The final state is therefore characterized by V_1, V_2, and S.

If now S, for different processes, was at times greater and at other times less than S^0, every neighboring state V_1, V_2, S would be accessible, for we have seen that the volumes may be altered freely. But this would violate Carathéodory's principle, because the process is adiabatic. Therefore, it must always be true that $S \geq S_0$ or that $S \leq S_0$.

If the final state is now made the initial state for an additional process, it follows from continuity considerations that the tendency of entropy to change must always be in the same direction. Whether S only increases or only decreases depends upon the sign of K in Equation (4.63). This is chosen so that the thermodynamic temperature is positive. From this choice follows the law of entropy increase. It also follows that for any change of state during which the value of the entropy does not remain constant, no adiabatic process can be found which will return the system from the final to the initial state. As a special case, no process in an *isolated* system may be found which can return the system to its original state. But by definition such processes, not reversible in every detail (see Chapter 1, Section 7), are irreversible, and this establishes Equation (4.2).

A few words are required concerning the conditions under which the proof given above pertaining to a system composed of two substances in thermal contact may be carried over to the treatment of complex systems. Obviously, the main point was the fact that the volumes were arbitrarily variable and that, in addition to these, there was only one other variable, the entropy. In general, it is true throughout all of thermodynamics that a system having n independent variables has $n - 1$ of these which possess geometric character and are arbitrarily variable, whereas only one thermal variable—temperature or entropy—occurs, for whose variability we have no intuition other than the principle of Carathéodory. Thus, the same argument may be advanced in connection with any thermodynamic system.

The principle of Thompson and the principle of Clausius, enunciated earlier in this chapter, are forms of the second law, and both assert the impossibility of doing something, namely the impossibility of transforming heat entirely into work without producing other changes in the system. The principle of Carathéodory, also a form of the second law, asserts the impossibility of doing something, in this case of establishing certain adiabatic processes. All forms of the second law which make immediate contact with physical intuition have this characteristic of asserting the impossibility of certain types of process.

12. Efficiency in Engineering Problems

Thus far, the efficiency η, introduced by Equation (4.21), has been used only for the advancement of the formalism, for example in the determination of a temperature scale. On the other hand, it is clear that engineering applications exist and that real heat engines may be treated; in which case η, as defined, measures the maximum efficiency of a heat engine.

By running the Carnot cycle in reverse, it is possible to withdraw heat from a reservoir at a lower temperature and deposit another quantity of heat at a higher temperature. In this case, the machine behaves as a "refrigerator" or "heat pump," depending upon whether the goal is the cooling of the low temperature source or the heating of the high temperature receptacle. In such instances, it is clear that the efficiency η defined by Equation (4.21) is inappropriate to the use at hand.

When refrigeration is the objective, one is interested in the amount of heat Δq_1 which can be *abstracted* from the low temperature reservoir by the expenditure of the work Dw. Thus, the most appropriate definition of efficiency is

$$\eta_R = \frac{\Delta q_1}{\Delta w}, \tag{4.65}$$

that is, the ratio of the heat abstracted to the work performed.

When the machine is utilized as a heat pump, one wishes to maximize the heat delivered to the hot reservoir for the expenditure of a given amount of work. Therefore, the appropriate efficiency is

$$\eta_P = \frac{\Delta q_2}{\Delta w}. \tag{4.66}$$

In keeping with the objective of this book, to discuss problem areas of understanding rather than to furnish an exhaustive treatment, we shall not pursue the topic of efficiency any further. The interested reader may consult any one of the numerous existing texts on thermodynamics.

13. The Helmholtz and Gibbs Free Energies

In Chapter 3, Section 6, attention was drawn to the importance, in thermodynamics, of possessing a large catalog of functions of state, and especially of having primary rather than derived functions. In that section, we synthesized the derived functions H, C_v, and C_p. With the new primary function S at our disposal, it is worthwhile to derive a few more functions.

A very useful function is the *Helmholtz* free energy A defined by

$$A = U - TS \tag{4.67}$$

(in which T is the thermodynamic temperature).

For a system of fixed mass we may use Equation (4.44) to express the work performed by the system during a change conducted *reversibly* and isothermally,

$$Dw = -dU + T\,dS. \tag{4.68}$$

For an isothermal process the differential of Equation (4.67) is

$$dA = dU - T\,dS. \tag{4.69}$$

Comparison of Equations (4.69) and (4.68) reveals that for an isothermal reversible process,

$$-dA = Dw, \tag{4.70}$$

that is, the decrease in A measures the reversible work performed *by* the system during the process. According to Equation (4.45), the maximum work available for a given change of state is performed when the change is conducted *reversibly*. Thus, the decrease in A measures the *maximum* work which the system can perform during the given change of state. For this reason, A is sometimes referred to as the *work content* of the system.

Another useful derived function of state is the Gibbs free energy denoted by the symbol G. This function is defined as follows:

$$G = U + pV - TS. \tag{4.71}$$

For a change of state in which both temperature and pressure are not changed (isothermal and isobaric), the change in G is

$$dG = dU + p\,dV - T\,dS. \tag{4.72}$$

Upon comparing Equation (4.72) with Equation (4.68) we see that

$$-dG = Dw - p\,dV. \tag{4.73}$$

Now for the reversible isothermal, isobaric change to which Dw refers, $p\,dV$ represents the volume work performed by the system. Hence, $Dw - p\,dV$ measures all the *reversible* work, except the volume work. Thus, Equation (4.73) shows that for a reversible, isothermal, isobaric process the decrease in G measures the amount of nonvolume work which the system can perform during the given change of state.

Whether or not we can prove that this is the *maximum* amount of nonvolume work depends upon whether $p\,dV$ has the same value whether the change is conducted reversibly or not.† By Equation (4.44) Dw is maximized but not necessarily the difference $Dw - p\,dV$. If the surroundings of the system relax so rapidly that a uniform hydrostatic pressure p may be defined during the conduct of the change of state (even when it is irreversible), the volume work will still be $p\,dV$. The situation is similar to that found in the discussion surrounding the identification of dH with the heat of change in Chapter 3, Section 6. From this argument it may be concluded that $Dw - p\,dV$ *is* maximized by conducting the process reversibly, provided that the environment relaxes rapidly. This follows from the fact that Dw is maximized and $p\,dV$ is the same, independent of whether or not the process is conducted reversibly. Many texts do not make this point clear and are usually content with the statement that, in general, the decrease in G measures the maximum amount of nonvolume work which can be drawn from an isothermal, isobaric process.

14. Legendre Transformations and Maxwell Relations

Consider a system consisting of a substance of constant mass, capable of performing volume work only. Such a system will possess two independent variables. The volume V may be chosen as one of these. If the system undergoes a reversible change of state, Equation (3.18) may be written in the form

$$dU = T\,dS - p\,dV, \tag{4.74}$$

where, for the reversible process, $T\,dS$ has been used to represent the heat absorbed and $p\,dV$ is the work performed. The quantities p and T, of course, only have thermodynamic significance in systems at equilibrium, and dS and dV define the elements of a quasistatic path which may be plotted in the macroscopic state space (Chapter 1, Section 5). Since dU itself is independent of whether the change of state is reversible or not, we may forget that $T\,dS$ and $p\,dV$ are the reversible heat and work, respectively; and Equation (4.74) may simply be regarded as an expression, in terms of the variables S and V, of the exact differential of U; that is, Equation (4.74) is of the form (2.1). By invoking Equation (2.4) we have

$$T = \left[\frac{\partial U}{\partial S}\right]_V, \tag{4.75}$$

$$p = -\left[\frac{\partial U}{\partial V}\right]_S. \tag{4.76}$$

† The author is unaware of any quantitative examples in which it can be shown that G does *not* measure the maximum nonvolume work available during a given change of state. Therefore, the contention that it does always measure the maximum work may be valid. On the other hand, he has never seen a general proof of this contention.

These equations represent examples of the data transformations which thermodynamics can provide. Equation (4.75) relates an equilibrium parameter T to the rate of change of an equilibrium parameter U with parameter S; V being held constant. In principle, measurements of U and S at constant V could therefore be used to measure T. Thus, the information on U and S can be transformed into information on T. Equation (4.76) is of a similar nature, except that here information concerning the equilibrium parameter p is obtained.

The important role of the part of the second law which postulates the existence of the entropy state function and prescribes a means for its measurement must be emphasized. Only through the use of Equation (4.38) (which depends upon this part of the law) have we been able to express dU in a form consistent with Equation (2.1). Without the second law this would be impossible. So much emphasis has been placed upon the part of the law which deals with *induced processes* (inequality (4.2)) that the importance of Equation (4.1) is sometimes obscured. Of course, Equation (4.74) makes combined use of *both* the first and second laws.

With Equation (4.74) at our disposal other transformations are immediately possible. For example, one may apply the reciprocity relation (2.6) to obtain

$$\left[\frac{\partial T}{\partial V}\right]_S = -\left[\frac{\partial p}{\partial S}\right]_V. \tag{4.77}$$

Relations such as (4.77) are known as Maxwell relations. Here the independent variables are S and V. It is natural to inquire into whether it is possible to develop Maxwell relations involving other sets of independent variables.

A powerful technique for accomplishing this is based upon the possibility of constructing derived functions of state which are combinations of established ones. For example, consider the enthalpy H defined by Equation (3.20). By taking the total differential of H we have

$$dH = dU + p\,dV + V\,dp. \tag{4.78}$$

Now substitute Equation (4.74) into Equation (4.78). This substitution may be regarded as the result of a well-known mathematical technique called a Legendre transformation.† The result is

$$dH = T\,dS + V\,dp, \tag{4.79}$$

from which it is obvious (since dH is a complete differential) that S and p are now being considered as the independent variables. Thus, it has been possible to switch (in a given system) from a means of development in which

† The Legendre transformation has found powerful application in classical mechanics. An excellent discussion of its mathematical significance may be found in H. B. Callen's *Thermodynamics* (John Wiley, 1960), pp. 90–98.

S and V had to be considered as the variables of state to one in which S and p enjoy the same privilege. Such is the nature of the Legendre transformation.

From Equation (4.79) the following equations (analogous to (4.75), (4.76), and (4.77)) are immediately derived:

$$T = \left[\frac{\partial H}{\partial S}\right]_p, \tag{4.80}$$

$$V = \left[\frac{\partial H}{\partial p}\right]_S, \tag{4.81}$$

$$\left[\frac{\partial T}{\partial p}\right]_S = \left[\frac{\partial V}{\partial S}\right]_p. \tag{4.82}$$

Legendre transformations may also be performed using the derived state functions A and G of Equations (4.67) and (4.71). Thus, proceeding in the same manner as with H we have for A

$$dA = dU - T\,dS - S\,dT, \tag{4.83}$$

into which we may substitute Equation (4.74). Then,

$$dA = -p\,dV - S\,dT, \tag{4.84}$$

so that the independent variables V and T are being considered. The following relations are derived immediately:

$$p = -\left[\frac{\partial A}{\partial V}\right]_T, \tag{4.85}$$

$$S = -\left[\frac{\partial A}{\partial T}\right]_V, \tag{4.86}$$

$$\left[\frac{\partial p}{\partial T}\right]_V = \left[\frac{\partial S}{\partial V}\right]_T. \tag{4.87}$$

In the case of G, we have

$$dG = dU + p\,dV + V\,dp - T\,dS - S\,dT. \tag{4.88}$$

After substituting Equation (4.74) this becomes

$$dG = V\,dp - S\,dT, \tag{4.89}$$

so that p and T are being treated as independent variables. From Equation (4.89) we obtain the following relations:

$$V = \left[\frac{\partial G}{\partial p}\right]_T, \tag{4.90}$$

$$S = -\left[\frac{\partial G}{\partial T}\right]_p, \tag{4.91}$$

$$\left[\frac{\partial V}{\partial T}\right]_p = -\left[\frac{\partial S}{\partial p}\right]_T. \tag{4.92}$$

V

Ideal Substances

1. Equation of State

It has been emphasized repeatedly that one of the most important applications of thermodynamics concerns the transformation of one set of empirical data into another. Such data constitute the clay to which the sculptor, "thermodynamics," applies itself. Nevertheless, it must be remembered that thermodynamic data are in a sense "extrathermodynamic"—pertaining to the *special* rather than the *general* properties of matter.

This extra-thermodynamic character must be re-emphasized when we consider hypothetical substances like *ideal gases* and *ideal solutions*. These concepts appear so frequently in the literature of thermodynamics that they may be misconstrued as inseparable adjuncts of the subject. Nothing could be further from the truth, and such ideal systems are merely substances whose behaviors yield certain special sets of thermodynamic data. No real substance is strictly ideal, and both the ideal gas and the ideal solution are chosen to represent the behaviors of real systems under certain limiting conditions. Since many real systems exist under conditions which are not far displaced from these limits, ideal behavior frequently represents a useful approximation.

The so-called ideal gas law is an example of an *equation of state*, and it is therefore appropriate to make some remarks concerning such equations. The term "equation of state" seems to have been given a variety of definitions (all more or less related) by different authors. We shall use it to denote the relation of a pair of *conjugate mechanical* variables such as pressure p and volume V to temperature T. The variables are "conjugate" in the sense that one is intensive and the other extensive, while their product has the dimensions of work or energy. Other examples of conjugate pairs are *surface tension* and *surface area*, and *electric potential* and *electric charge*. Sometimes a system may possess more than two independent variables, in which case the equation of state will relate the two conjugate mechanical variables to the temperature and whatever other independent variables must be considered; but the underlying theme still remains the connection between conjugate *mechanical* variables and the *thermal* variable temperature.

The term "equation of state" is somewhat misleading, for we shall see that the equilibrium behavior of a substance is not completely described by its equation of state.

2. The Ideal Gas

The equation of state for the ideal gas is so well-known that it hardly needs comment. On the other hand, a few points are worthy of exposure, and for this reason we shall devote some space to it. The equation is usually written in the form

$$pV = nRT, \tag{5.1}$$

where R is the gas constant per mole, n represents the number of moles of gas contained within the system, and T is the ideal gas or Kelvin temperature.

When p is measured in atmospheres and V in cubic centimeters, $R = 82.056 \pm 0.005$ cc atm per degree Kelvin per mole.

Equation (5.1) is a limiting form, towards which the equations of state of real gases are observed to converge when their densities $\left(\text{measured by } \frac{n}{V}\right)$ approach zero. As a matter of fact, Equation (5.1) is not a bad approximation for many gases at reasonably finite densities; for example, air at conventional pressures and temperatures. In general, the equation is useful out to higher densities if the temperature is higher.

The appearance of T as we find it in Equation (5.1) is the result of somewhat circular reasoning, for the equation of state of an ideal gas *is often used* in the definition of T. On the other hand, we saw in Chapter 4 how it was possible to use the second law to establish a *thermodynamic* temperature which is independent of any special substance; and it is of interest, therefore, to see how the Kelvin scale is related to it. The empirical temperature θ of Chapter 3 could be established in terms of the measurable properties of some real substance (the thermometer). For example, at constant pressure one might employ the equilibrium volume of a solid, liquid, or gas (assuming the absence of chemical transformation) to specify temperature.

Thus, the volume of a given substance in equilibrium with ice melting in air under a pressure of one atmosphere could be chosen as the zero of temperature, while the volume attained when in equilibrium with water boiling at one atmosphere may be arbitrarily chosen to represent 100°. The increase of volume between these two points may be subdivided uniformly into 100 equal intervals, each of which may be used to define a degree.

It is clear that this method of subdivision permits specific intermediate readings, obtained by reference to two *different* substances, to correspond to two different empirical temperatures even though they may have the same numerical value. For example, 50° by one method of reckoning may not correspond to 50° determined through application of another method. Stated in another way, two thermometers, both registering 50°, may not be in thermal equilibrium with one another; that is, they may not satisfy Equation (3.1). The numerical specification of the temperature scale will depend, intimately, upon the manner in which the volume of the thermometric substance depends upon empirical temperature. The only exceptions will be the fiducial points at 0 and 100° which will have to be the same for all scales.

It is observed that when highly attenuated gases are employed (in this manner) as thermometric substances, two such gases which yield the same numerical value of temperature are almost in thermal equilibrium when placed in contact with one another. This suggests that the temperature dependences of their volumes are very nearly identical, and leads to the concept of the ideal gas as the natural limiting condition for all gases as their densities tend toward zero.

In effect it is discovered, through extrapolation of the behaviors of many gases to zero density, that the gas volume at the boiling point of water V_{100} (where the subscript 100 refers to 100°) bears the following relation to V_0, the volume at the melting point of ice under one atmosphere of pressure:

$$V_{100} = V_0 + (100/273.15)V_0. \tag{5.2}$$

It will be recognized that it is the familiar centrigrade scale which is determined by the melting point of ice and the boiling point of water, and it is clear from the appearance of the awkward ratio (100/273.15) in Equation (5.2) that this scale is not convenient when a highly attenuated (ideal) gas is employed as a thermometer.

If one simply plots the volume V of a highly attenuated gas versus temperature in the range between 0 and 100°C, the result—by definition—must be a straight line, for the gas volume was employed in the first place to establish the subdivisions of the scale. The equation of this line will be (by virtue of Equation (5.2))

$$V = V_0 + (T'/273.15)V_0, \tag{5.3}$$

where T' is the centigrade temperature. The straight (V against T') line may be extrapolated until it intercepts the temperature axis, at which point V will be zero. By calling the associated temperature T_0', we have from Equation (5.3)

$$O = V_0 + (T_0'/273.15)V_0 \tag{5.4}$$

or

$$T_0' = -273.15.$$

This result demonstrates that a new temperature T, such that

$$T = T' + 273.15, \tag{5.5}$$

will have the value zero at the point where the extrapolated volume of the ideal gas vanishes. T is the Kelvin temperature and is symbolized by °K. By symmetry it is well suited to the ideal gas thermometer. Its appearance in Equation (5.1), in just the way it does, is a direct result of the use of the ideal gas in the *definition* of the Kelvin scale and therefore invokes some elements of circular reasoning.

It must be borne in mind that any real gas will have ceased to be ideal (even if highly attenuated) at $T = 0$°K, sometimes called absolute zero. This poses no problem because the ideal gas is an abstraction in the first place, chosen to match the limiting behavior of real gases in some range of temperature. One must be careful not to read too much into the expression "absolute zero," especially when it is defined by the ideal gas scale. We are at liberty to choose θ, the empirical temperature, in a variety of ways, and might just as well have chosen $\theta = \ln T$ as $\theta = T$, in which case at $T = 0$, θ would be $-\infty$, and the temperature scale would be bottomless.

On the other hand, the Kelvin scale can be invested with more fundamental importance by exhibiting its relation to the thermodynamic scale. This is accomplished in the next section.

3. Internal Energy of the Ideal Gas and Relation between the Kelvin and Thermodynamic Scales

Equation (4.74) may be treated by the method of Equation (2.22) to yield

$$\left[\frac{\partial U}{\partial V}\right]_T = T\left[\frac{\partial S}{\partial V}\right]_T - p. \tag{5.6}$$

This may be further transformed through the use of Equation (4.87) to yield

$$\left[\frac{\partial U}{\partial V}\right]_T = T\left[\frac{\partial p}{\partial T}\right]_V - p. \tag{5.7}$$

If Equation (5.7) is applied to an ideal gas, Equation (5.1) may be substituted into its right side with the result

$$\left[\frac{\partial U}{\partial V}\right]_T = 0. \tag{5.8}$$

This means that the internal energy of an ideal gas is independent of the volume and depends only upon temperature. But how does it depend upon temperature? We shall discuss this matter later. In the meantime, it is possible to illustrate the connection between the Kelvin and thermodynamic scales using only Equation (5.8), and therefore only Equation (5.1).

To accomplish this, carry one mole ($n = 1$) of an ideal gas around the reversible cycle illustrated in Figure (4.2). For the isothermal legs of the cycle, conducted now at Kelvin temperatures T_2 and T_1, respectively, Equation (5.8) assures us that $\Delta U = 0$. Therefore, Equation (3.18) yields

$$\Delta q_2 = \Delta w_2 = \int_{V_A}^{V_B} p\, dV, \tag{5.9}$$

$$\Delta q_1 = \Delta w_1 = \int_{V_C}^{V_D} p\, dV, \tag{5.10}$$

where Δq_2, Δw_2, Δq_1, Δw_1, and V_A, V_B, V_C, and V_D have the same meanings as in Figure 4.2, and the terms on the extreme right of both Equations (5.9) and (5.10) arise because the ideal gas is capable only of volume work. Upon substituting Equation (5.1) with $n = 1$ into the integrals on the right, we have

$$\Delta q_2 = \Delta w_2 = RT_2 \ln \frac{V_B}{V_A}, \tag{5.11}$$

$$\Delta q_1 = \Delta w_1 = RT_1 \ln \frac{V_D}{V_C}. \tag{5.12}$$

Next consider the adiabatic legs of the cycle in Figure 4.2. Since now $Dq = 0$, Equation (3.18) yields

$$\begin{aligned} U_C - U_B &= -\Delta w_{21}, \\ U_A - U_D &= -\Delta w_{12}, \end{aligned} \tag{5.13}$$

where U_A, U_B, U_C, and U_D are the internal energies at the points A, B, C, and D. But points A and B are both at temperature T_2 and C and D are both at T_1; therefore, since U depends only on temperature,

$$\begin{aligned} U_A &= U_B, \\ U_C &= U_D. \end{aligned} \tag{5.14}$$

If we consider a differential element of one of the adiabatic legs, Equation (3.18) requires

$$dU = -Dw = -p\, dV. \tag{5.15}$$

By substituting from Equation (5.1) we get

$$dU = -RT\, d \ln V$$

or

$$\frac{dU}{T} = -R\, d \ln V. \tag{5.16}$$

Integration of the last form of Equation (5.16) between B and C and between

D and A over both adiabatic legs of the cycle yields

$$\int_{T_2}^{T_1} \frac{dU}{T} = -R \ln \left[\frac{V_C}{V_B}\right], \tag{5.17}$$

$$\int_{T_1}^{T_2} \frac{dU}{T} = -R \ln \left[\frac{V_A}{V_D}\right]. \tag{5.18}$$

Now the integrands on the left of Equation (5.17) and (5.18) depend only upon temperature (since U depends only upon temperature), and because the limits are reversed they are the negatives of each other. The same must be true of the right-hand sides of these equations so that

$$\frac{V_D}{V_A} = \frac{V_C}{V_B} \quad \text{or} \quad \frac{V_B}{V_A} = \frac{V_C}{V_D}. \tag{5.19}$$

Now the efficiency defined by Equation (4.21) is

$$\eta = \frac{\Delta w}{\Delta q_2} = \frac{\Delta w_2 + \Delta w_{21} + \Delta w_1 + \Delta w_{12}}{\Delta q_2}$$

$$= \frac{RT_2 \ln \left[\frac{V_B}{V_A}\right] + U_B - U_C + RT_1 \ln \left[\frac{V_D}{V_C}\right] + U_D - U_A}{RT_2 \ln \left[\frac{V_B}{V_A}\right]}$$

$$= \frac{T_2 - T_1}{T_2}, \tag{5.20}$$

where Equations (5.14) and (5.19) have been used to achieve the final reduction. Comparison of this result with Equation (4.24) shows that T must be proportional to the thermodynamic temperature. Finally, because the thermodynamic scale is determined by arbitrarily fixing, at 273.15 degrees, the temperature at the melting point of ice under one atmosphere of pressure, the proportionality constant must be unity; and the thermodynamic and Kelvin scales must, therefore, be identical. Upon returning to Equation (4.74) and applying the method of Equation (2.22), we can derive

$$\left[\frac{\partial U}{\partial T}\right]_V = T\left[\frac{\partial S}{\partial T}\right]_V. \tag{5.21}$$

Now inspection reveals that it is not possible to transform Equation (5.21) in a manner similar to that in which Equation (5.6) was transformed into Equation (5.7) so that the right-hand side may be evaluated by use of the ideal gas law (5.1). Thus, the dependence of U upon T cannot be determined from Equation (5.1) alone; and therefore (5.1) does not determine everything about the *state* of an ideal gas. It is, therefore, somewhat of a misnomer to refer to it as the equation of state. We may complete the description of an ideal gas by specifying how its internal energy depends upon temperature.

To accomplish this in a manner consistent with the meaning of Equation (5.1), one in effect observes U and T in real gases in the limit of vanishing density. The quantity on the left of Equation (5.21) is C_v, the *heat capacity at constant volume* (see Equation (3.19)). This is to be compared with C_p the *heat capacity at constant pressure* defined in Equation (3.23).

In principle, C_v may be measured in the following manner. Work is performed on a substance in an adiabatic environment, and the change dU in its internal energy is established through the use of Equation (3.14). At the same time, having defined the temperature scale, dT may be measured. If in addition to the requirement of adiabaticity the volume of the system is maintained constant during the performance of this work, the ratio of dU to dT will determine the partial derivative $[\partial U/\partial T]_V = C_v$. Notice that the work performed on the system cannot be volume work since V is held constant. It may, for example, involve stirring.

For a true ideal gas we have shown that Equation (5.1) requires U to be independent of V. As a result, the measurement of C_v for a substance obeying the relation (5.1) does not require the maintenance of constant volume, for

$$\left[\frac{\partial U}{\partial T}\right]_V = \frac{dU(T)}{dT} = C_v. \tag{5.22}$$

Measurements of C_v in attenuated monatomic gases suggest that in the limit of zero density

$$C_v \rightarrow \text{const.}\dagger \tag{5.23}$$

For a monatomic gas the following relation is in fact suggested.

$$C_v = \tfrac{3}{2}nR. \tag{5.24}$$

Integration of Equation (5.22) with Equation (5.23) substituted yields

$$U = C_vT, \tag{5.25}$$

where the constant of integration has been set arbitrarily equal to zero.

Equations (5.1) and (5.25), together, do in fact determine the complete behavior of the ideal gas in the sense that all further thermodynamic quantities may be evaluated for the gas by thermodynamic reasoning alone. For example, in the next section we use these relations to determine the entropy of an ideal gas.

† At high enough temperatures the same relation apparently holds for polyatomic gases although c_v possesses a larger value than that shown in Equation (5.24).

4. Variation of the Entropy of an Ideal Gas

Substitution of Equation (5.24) into Equation (3.19), and of the latter into Equation (5.21), yields

$$\left[\frac{\partial S}{\partial T}\right]_V = \frac{C_v}{T}. \tag{5.26}$$

Furthermore, substitution of Equation (5.1) into Equation (4.87) results in the relation

$$\left[\frac{\partial S}{\partial V}\right]_T = \frac{nR}{V}. \tag{5.27}$$

If T and V are chosen as independent variables, one may express the differential of the entropy (in accordance with Equation (2.1)) in the form

$$dS = \left[\frac{\partial S}{\partial T}\right]_V dT + \left[\frac{\partial S}{\partial V}\right]_T dV. \tag{5.28}$$

Substitution of Equations (5.26) and (5.27) into Equation (5.28) yields

$$dS = C_v\, d\ln T + nR\, d\ln V. \tag{5.29}$$

Suppose the gas is carried from the thermodynamic state 1, characterized by T_1, V_1, to state 2, characterized by T_2, V_2. Since the change in entropy $S_2 - S_1$ is independent of the path (since S is a function of state), we may choose the following convenient route in state space. First, the temperature is changed from T_1 to T_2 while volume remains fixed at V_1. Then volume is changed from V_1 to V_2 while the temperature remains fixed at T_2. Thus, the passage from state 1 to 2 is effected in two steps:

$$\begin{aligned} &\text{(a)} \quad (T_1, V_1) \rightarrow (T_2, V_1), \\ &\text{(b)} \quad (T_2, V_1) \rightarrow (T_2, V_2). \end{aligned} \tag{5.30}$$

In step (a), $d\ln V = 0$ so that Equation (5.29) may be simplified to

$$dS = C_v\, d\ln T. \tag{5.31}$$

The entropy change during this step may be denoted by ΔS_a, and may be evaluated by direct integration of Equation (5.31) between T_1 and T_2. Thus,

$$\Delta S_a = C_v \ln\left[\frac{T_2}{T_1}\right]. \tag{5.32}$$

During step (b), $d\ln T = 0$ so that Equation (5.29) becomes

$$dS = nR\, d\ln V. \tag{5.33}$$

The entropy change during this step may be symbolized by ΔS_b and may be evaluated by the direct integration of Equation (5.33) between V_1 and V_2

$$\Delta S_b = nR \ln \left[\frac{V_2}{V_1}\right]. \tag{5.34}$$

Now the total entropy change $S_2 - S_1$ is merely the sum of the changes for the two steps connecting states 1 and 2,

$$S_2 - S_1 = \Delta S_a + \Delta S_b, \tag{5.35}$$

and using Equations (5.32) and (5.34) gives

$$S_2 - S_1 = nR \ln \left[\frac{V_2}{V_1}\right]\left[\frac{T_2}{T_1}\right]^{C_v/nR}. \tag{5.36}$$

For a reversible adiabatic expansion between states 1 and 2, the volume-temperature relationship may be determined by application of Equation (5.36). For a reversible adiabatic process, according to Equation (4.38),

$$Dq = 0 = T\,dS \quad \text{or} \quad dS = 0. \tag{5.37}$$

Thus, $S_2 - S_1$ in Equation (5.25) may be set equal to zero with the result,

$$\left[\frac{T_2}{T_1}\right] = \left[\frac{V_1}{V_2}\right]^{nR/C_v}. \tag{5.38}$$

This is the desired volume-temperature relation.

5. The Entropy of Mixing of Two Ideal Gases

Experiments on mixtures of highly attenuated gases have indicated that the pressure of the mixture is approximately equal to the sum of the pressures which each component would exert by itself were it the only gas present. The pressure which each component exerts by itself is called its *partial pressure*; and it appears as though, in the limit of vanishing density, the relation enunciated above becomes exact; that is, the pressure of the mixture is precisely the sum of the partial pressures. Thus, we adopt as another characteristic of ideal gas behavior (this time for mixtures of ideal gases) the relation

$$p = \sum_i p_i, \tag{5.39}$$

where p is the pressure of the mixture and p_i is the partial pressure of the ith component. This is the famous Dalton's law of partial pressures.

Like the ideal gas law itself and the assignment (5.25) for the internal energy of an ideal gas, Dalton's law is extra-thermodynamic. It is empirical to the extent that it represents the limiting behavior of a mixture of real gases.

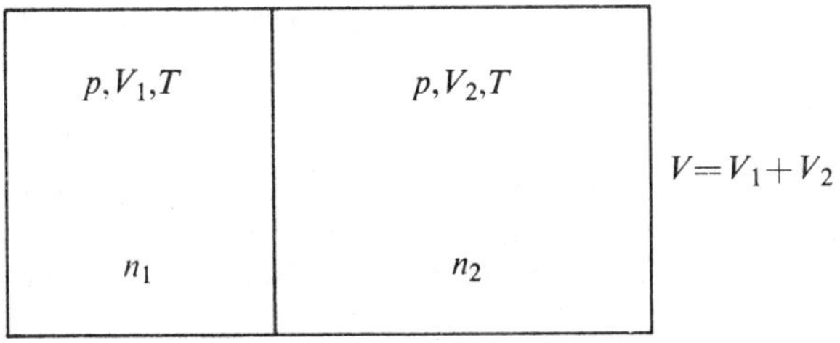

FIGURE 5.1. *Two gases ready to be mixed.*

On an empirical basis Equation (5.39) does suggest that several ideal gases contained within the same volume are not aware of each other's presence. They each go right on exerting their own pressures independent of whether another gas is present or not.

Consider two different ideal gases 1 and 2 contained initially as in Figure 5.1 in separate compartments having the volumes V_1 and V_2. Both gases are at the same temperature T and pressure p. The ideal gas law (Equation (5.1)) then requires the number of moles of each to be

$$n_1 = \frac{pV_1}{RT}, \qquad n_2 = \frac{pV_2}{RT}, \tag{5.40}$$

respectively.

If the partition between the two compartments is removed, the gases will mix and each will fill the entire volume V. It is assumed that the temperature is maintained at T. If the gases are indifferent to one another's presence, the new pressure of each may be computed from Equation (5.1), assuming the volume of each to be

$$V = V_1 + V_2. \tag{5.41}$$

Thus,

$$p_1 = \frac{n_1RT}{V}, \qquad p_2 = \frac{n_2RT}{V}. \tag{5.42}$$

The total pressure of the mixture given by Equation (5.39) is

$$p_1 + p_2 = \frac{RT}{V}(n_1 + n_2) = \frac{RT}{V}\left\{\frac{p}{RT}(V_1 + V_2)\right\} = p, \tag{5.43}$$

where both Equations (5.40) and (5.41) have been used. Thus, if the original pressures in each compartment were identical, the final total pressure after mixing will be the same as the initial pressures, it being understood that the temperature remains fixed at T.

If the gases are indifferent to one another's presence, the process of mixing is a process of isothermal expansion viewed from the standpoint of either gas alone. The entropy increase during such an isothermal expansion may be computed by performing the expansion reversibly and computing the heat Dq absorbed. According to Equation (4.38), $\frac{Dq}{T}$ then represents the entropy change. Now the heat Dq absorbed by an ideal gas during an isothermal expansion has the form given by Equation (5.11) or (5.12) multiplied by n, the number of moles, since Equations (5.11) and (5.12) refer to one mole of gas. Thus, for gas 1

$$\Delta S_1 = n_1 R \ln \left[\frac{V}{V_1}\right], \tag{5.44}$$

and for gas 2

$$\Delta S_2 = n_2 R \ln \left[\frac{V}{V_2}\right], \tag{5.45}$$

where ΔS_1 and ΔS_2 are the corresponding entropy changes. The total change of entropy which attends the mixing process is thus

$$\begin{aligned} \Delta S = \Delta S_1 + \Delta S_2 &= n_1 R \ln \left[\frac{V}{V_1}\right] + n_2 R \ln \left[\frac{V}{V_2}\right] \\ &= -n_1 R \ln \left[\frac{n_1}{n_1 + n_2}\right] - n_2 R \ln \left[\frac{n_2}{n_1 + n_2}\right] \\ &= -n_1 R \ln x_1 - n_2 R \ln x_2, \end{aligned} \tag{5.46}$$

where Equations (5.40) and (5.41) have been used and

$$x_1 = \frac{n_1}{n_1 + n_2}$$

and

$$x_2 = \frac{n_2}{n_1 + n_2} \tag{5.47}$$

are the *mole fractions* in the mixture of gases 1 and 2, respectively. ΔS defined by Equation (5.46) is known as the *entropy of mixing* of two ideal gases. Notice that during the mixing process temperature, pressure, and total volume remain invariant.

Since the internal energy of each indifferent ideal component is independent of volume, the mixing process does not involve any change in energy; each component merely undergoes an isothermal expansion. Furthermore, since the volume remains constant and a constant pressure change is involved, there is no *enthalpy of mixing*.

In closing this section, it must be emphasized again that Equation (5.46) cannot be derived from Equations (5.1) and (5.25) by thermodynamic reasoning alone. Extra-thermodynamic information is required. This is furnished by the assumption (suggested by Dalton's law) that the components of an ideal gas mixture are mutually indifferent.

6. Ideal Solutions

The ideal gas mixture considered in the previous section furnishes an example of a solution, albeit a gaseous solution. Since it is composed of ideal gases, it may be termed an *ideal solution*. As a solution it is characterized by the fact that its volume equals the sum of the volumes of the components (see Equation (5.41)), the pressure being held constant during mixing. Furthermore, there is no *enthalpy of solution*. Finally, mixing is attended by an entropy change specified by Equation (5.46).

Now it has been observed† that some liquid and solid solutions exhibit (at least approximately) these same mixing characteristics; namely, that upon mixing at constant temperature and pressure there is:

(1) no change of volume;
(2) no enthalpy of mixing;
(3) an entropy change specified by Equation (5.46).

Consequently, thermodynamicists have been prompted to call any solution (gas, liquid, or solid) which conforms to these three requirements an *ideal solution*. Actually, it may be shown (and we shall do so below) that the three requirements are not independent, and that the first is demanded whenever the last two apply. Thus, it is sufficient to define an ideal solution by the requirements that upon preparation from its pure components (under constant temperature, constant pressure conditions) there shall be:

(1) an entropy change specified by Equation (5.46) (in the mole fraction form),
(2) no enthalpy of mixing.

These conditions may be presented in more quantitative form as follows. The total enthalpy of the collection of pure components, there being n_i moles of species i, is

$$H^0 = \sum n_i H_i^0, \tag{5.48}$$

† The entropy change is not measured directly but is inferred from the measurement of relationships derived as consequences of Equation (5.46). See Chapter 9, Sections 2 through 5.

where H_i^0 is the enthalpy of a mole of pure species i. The enthalpy of the solution prepared from this collection of pure components is given by Equation (3.36),

$$H = \sum_i n_i \bar{H}_i. \tag{5.49}$$

Thus, the heat absorbed upon solution is

$$\Delta H = H - H^0 = \sum_i n_i(\bar{H}_i - H_i^0). \tag{5.50}$$

The volume change accompanying the process of solution may be expressed in an entirely analogous manner as

$$\Delta V = V - V^0 = \sum_i n_i(\bar{V}_i - V_i^0). \tag{5.51}$$

Conditions (1) and (2) above, defining an ideal solution, may now be expressed quantitatively using Equations (5.46) and (5.50). Thus,

$$(1) \quad (\Delta S)_{T,p} = -R \sum_i n_i \ln x_i, \tag{5.52}$$

and

$$(2) \quad (\Delta H)_{T,p} = \sum_i n_i(H_i^0 - \bar{H}_i) = 0, \tag{5.53}$$

where the subscripts T and p are meant to indicate isothermal and isobaric (constant pressure) conditions.

Although we do not intend, in this book, to rely upon any nonmacroscopic arguments, the special substances which have been discussed in this chapter possess characteristics which cannot be demonstrated by thermodynamic reasoning alone. Therefore, it is only appropriate to make a few remarks concerning the molecular theory basis of both ideal gases and ideal solutions. But it should be borne in mind that this explanation of mechanism is a nonthermodynamic interlude which merely justifies the experimental results embodied in Equations (5.1) and (5.25).

Ideal gases are substances whose individual molecules do not interact with one another (except with vanishingly small collision frequencies so that equilibrium may be established). This is why real gases behave ideally in the limit of zero density. Their molecules are then so far apart that on the average they exert no forces upon one another. Furthermore, the absence of molecular interaction explains why the internal energy of an ideal gas is independent of its volume. A change in the average distance between molecules does not change the potential energy of interaction since there is none to begin with. Finally, Dalton's law of partial pressures is clearly to be expected when the molecules are indifferent to one another's presence, that is, when they do not interact.

The ideal solution represents a somewhat different situation. Here there may be strong interactions between the molecules, but they are nonspecific.

For example, if a solution consists of two species A and B (a *binary* solution), then the interaction between two A molecules is the same as that between two B molecules, or between an A and a B.

It is a simple step intuitively to see that the absence, in ideal solutions, of both heats and volumes of mixing is a reasonable consequence of such nonspecificity of forces. Thus, energetically nothing is changed upon mixing, and so it is not surprising that there is no enthalpy of mixing. Furthermore, since forces are nonspecific, nothing in the character of the cohesion is changed by mixing, so it is not surprising that the volume change on mixing is zero.

In regard to actual substances which are apt to form solutions whose behaviors are approximately ideal, it is to be expected that components with physically similar molecules are most likely to do so. For example, mixtures of isotopes form almost perfect ideal solutions. Furthermore, mixtures of molecules like toluene whose molecular structure is

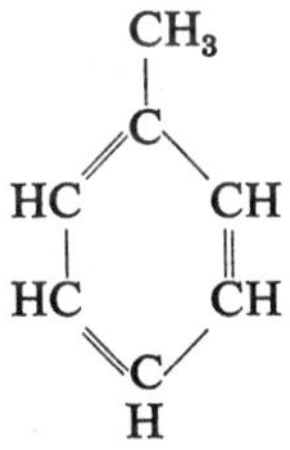

and benzene with the structure

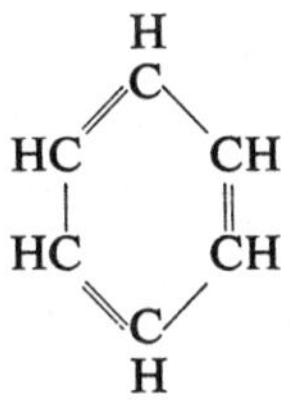

form solutions which are almost ideal.

An *ideal gas solution* is a special limiting case—*no* interaction between A and B is the same as *no* interaction between A and A or between B and B.

7. $\bar{G}_i$ for a Component of an Ideal Solution and Proof that the Volume of Mixing is Zero

In the following chapters we shall make repeated use of an expression which exhibits the dependence of $\bar{G}_i$, the partial molar Gibbs free energy of the ith species in an ideal solution, on the composition of the solution.

The expression for $\bar{G}_i$ will now be derived from Equations (5.52) and (5.53), the defining characteristics of an ideal solution. Along the way we shall prove that the volume of mixing is zero, a condition which may be expressed quantitatively through the use of Equation (5.51). Thus,

$$(\Delta V)_{T,p} = \sum_i n_i(\bar{V}_i - V_i^0) = 0. \tag{5.54}$$

To begin with, the Gibbs free energy of mixing may also be expressed in a form like Equation (5.50) or Equation (5.51),

$$(\Delta G)_{T,p} = \sum_i n_i(\bar{G}_i - G_i^0), \tag{5.55}$$

where G_i^0 is the Gibbs free energy of a mole of pure species i. At this juncture, it is important to note that G_i^0, since it refers to a pure component, does not depend upon the composition of the solution but only upon T and p. Thus, we may write

$$G_i^0 = G_i^0(T,p). \tag{5.56}$$

From the definitions of H and G (Equations (3.20) and (4.71)) it is apparent that

$$G = H - TS, \tag{5.57}$$

and that for a *constant temperature process*

$$(\Delta G)_{T,p} = \Delta H - T\,\Delta S. \tag{5.58}$$

In particular, for the mixing process,

$$(\Delta G)_{T,p} = -T\,\Delta S, \tag{5.59}$$

where Equation (5.53) has been used.

Substitution from Equations (5.52) and (5.55) into Equation (5.59) yields

$$\begin{aligned} \sum_i n_i(\bar{G}_i - G_i^0) &= RT\sum_i n_i \ln x_i \\ &= RT\sum_i n_i \ln n_i - RT\sum_i n_i \ln\left[\sum_k n_k\right], \end{aligned} \tag{5.60}$$

where the definition

$$x_i = n_i/\sum_k n_k \tag{5.61}$$

has been used. By differentiating Equation (5.60) with respect to n_j with T and p and all other mole numbers held constant we get

$$\begin{aligned} &\sum_i n_i\left[\frac{\partial \bar{G}_i}{\partial n_j}\right]_{T,p} + \bar{G}_j - G_j^0 \\ &\qquad = RT + R\ln n_j - RT\ln \sum_k n_k - RT\sum_i n_i/\sum_k n_k. \end{aligned} \tag{5.62}$$

Now with temperature and pressure constant, a relation of the kind (3.35) which holds for $\bar{G}_i$ as well as $\bar{V}_i$ (and in fact for any partial molar quantity) requires the first term on the left to vanish. Furthermore, the last term on the right is

$$-RT, \tag{5.63}$$

which cancels the first term on the right. The remaining terms on the right may be combined to yield

$$RT \ln x_j. \tag{5.64}$$

Thus, Equation (5.62) becomes

$$\bar{G}_j = G_j^0(T,p) + RT \ln x_j. \tag{5.65}$$

This is the characteristic form for the partial molar, Gibbs free energy of a component of an ideal solution.

To prove Equation (5.54) (that the volume change upon mixing is zero) we proceed as follows. First, we note that the following set of successive transformations is valid:

$$\begin{aligned}\left[\frac{\partial \bar{G}_i}{\partial p}\right]_{T,n} &= \left\{\frac{\partial}{\partial p}\left[\frac{\partial G}{\partial n_i}\right]_{T,p}\right\}_{T,n} \\ &= \left\{\frac{\partial}{\partial n_i}\left[\frac{\partial G}{\partial p}\right]_{T,n}\right\}_{T,p} = \left[\frac{\partial V}{\partial n_i}\right]_{T,p} = \bar{V}_i.\end{aligned} \tag{5.66}$$

In these equations, the subscript n in $[\ ]_{T,n}$ is an abbreviation for $n_1, n_2, \ldots,$ the mole numbers of all of the components in the solution; whereas, in $[\ ]_{T,p}$ the subscripts should also contain all $n_j \neq n_i$. We have omitted these symbols in $[\ ]_{T,p}$ in the interest of notational simplicity. No confusion should result from this since it will always be clear when the omission has been made; this practice will be followed from now on. Proceeding from left to right in Equation (5.66), the second term is obtained from the first through the use of the definition (3.29) of a partial molar quantity, while the third term is derived from the second through an inversion of the order of differentiation. The fourth term follows from the third through application of the relation (4.90), and the fifth term is derived from the fourth by the use of the definition (3.29). In passing, we might remark that by a similar series of transformations, using Equation (4.91) instead of Equation (4.90), one can show that

$$\left[\frac{\partial \bar{G}_i}{\partial T}\right]_{p,n} = -\bar{S}_i. \tag{5.67}$$

Upon returning to the proof of Equation (5.54), it follows from Equations (5.65) and (5.66) that

$$\bar{V}_i = \left[\frac{\partial \bar{G}_i}{\partial p}\right]_{T,n} = \left[\frac{\partial G_i^0}{\partial p}\right]_{T,n}, \tag{5.68}$$

since $RT \ln x_i$ is independent of p. But Equation (5.66) may equally well be applied to a mole of pure component i with the result

$$V_i^0 = \left[\frac{\partial G_i^0}{\partial p}\right]_{T,n}. \tag{5.69}$$

Comparison of Equation (5.69) with Equation (5.68) reveals that for an ideal solution,

$$\bar{V}_i = V_i^0, \tag{5.70}$$

and substitution of this relation in Equation (5.51) gives

$$(\Delta V)_{T,p} = \sum_i n_i(\bar{V}_i - V_i^0) = \sum n_i(V_i^0 - V_i^0) = 0, \tag{5.71}$$

which is Equation (5.54). Thus, the volume of mixing for an ideal solution is zero.

VI

Some Useful Formulas

1. Compressibility and Expansivity and the Relation between C_v and C_p

We have made the point that one of the major contributions of thermodynamics is concerned with the derivation of formulas interconnecting measurable macroscopic variables. In this chapter, we shall employ thermodynamic methods to derive several such formulas. These derivations are presented, not only because the formulas are useful but because they illustrate the technique. We begin by introducing two new derived quantities, the *thermal expansivity* α and the *isothermal compressibility* β. The expansivity is defined (for a simple two-variable system) as

$$\alpha = \frac{1}{V}\left[\frac{\partial V}{\partial T}\right]_p, \tag{6.1}$$

and the compressibility is

$$\beta = -\frac{1}{V}\left[\frac{\partial V}{\partial p}\right]_T. \tag{6.2}$$

It is clear from their formulations that α and β measure the fractional changes in volume which attend changes in temperature and pressure, respectively.

Now it may be shown that α and β appear in a formula which connects C_p, the heat capacity at constant pressure with C_v, the heat capacity at constant volume. The proof proceeds as follows.

Since Dq in Equation (4.15) refers to a reversible change, we may substitute $T\,dS$ for it. If, furthermore, we replace θ in Equation (4.15) by T (now that the ideal gas temperature has been defined and identified with the thermodynamic temperature), the equation becomes

$$T\,dS = \left[\frac{\partial U}{\partial T}\right]_V dT + \left\{p + \left[\frac{\partial U}{\partial V}\right]_T\right\} dV. \tag{6.3}$$

After applying Equation (2.32) to Equation (6.3) to obtain the derivative of S with respect to T at constant p, and employing the definition of C_v given in Equation (3.19), this becomes

$$T\left[\frac{\partial S}{\partial T}\right]_p = C_v + \left\{p + \left[\frac{\partial U}{\partial V}\right]_T\right\}\left[\frac{\partial V}{\partial T}\right]_p. \tag{6.4}$$

By applying Equation (2.32) again, this time to Equation (4.74), we obtain the derivative of U with respect to V, holding T constant,

$$\left[\frac{\partial U}{\partial V}\right]_T = T\left[\frac{\partial S}{\partial V}\right]_T - p. \tag{6.5}$$

This may be substituted into the bracketed expression in Equation (6.4) to yield

$$T\left[\frac{\partial S}{\partial T}\right]_p = C_v + T\left[\frac{\partial S}{\partial V}\right]_T\left[\frac{\partial V}{\partial T}\right]_p. \tag{6.6}$$

By using Equation (2.32) once again, this time in connection with Equation (4.79), we get the derivative of H with respect to T with p held constant

$$\left[\frac{\partial H}{\partial T}\right]_p = T\left[\frac{\partial S}{\partial T}\right]_p = C_p, \tag{6.7}$$

where the definition of C_p given in Equation (3.23) has been employed. The equation on the right of Equation (6.7) may now be substituted into Equation (6.6) to replace $T\left[\frac{\partial T}{\partial S}\right]_p$ with C_p. If at the same time Equation (4.87) is substituted into the right side of Equation (6.6), the result is

$$C_p - C_v = T\left[\frac{\partial p}{\partial T}\right]_V\left[\frac{\partial V}{\partial T}\right]_p = VT\left[\frac{\partial p}{\partial T}\right]_V\left\{\frac{1}{V}\left[\frac{\partial V}{\partial T}\right]_p\right\} = VT\alpha\left[\frac{\partial p}{\partial T}\right]_V, \tag{6.8}$$

where the definition (6.1) has been employed. This result still requires further development. We may accomplish this by applying Equation (2.26) to obtain

$$\left[\frac{\partial p}{\partial T}\right]_V = -\left[\frac{\partial p}{\partial V}\right]_T\left[\frac{\partial V}{\partial T}\right]_p = -\frac{1}{V}\left[\frac{\partial V}{\partial T}\right]_p \Big/ \frac{1}{V}\left[\frac{\partial V}{\partial p}\right]_T = \frac{\alpha}{\beta}, \tag{6.9}$$

where the definitions (6.1) and (6.2) have been used. Substitution of Equation (6.9) into Equation (6.8) yields the sought-for result

$$C_p - C_v = \frac{VT\alpha^2}{\beta}. \tag{6.10}$$

This general result may be examined for the special case of the ideal gas. According to Equations (5.25) and (5.1), for one mole of ideal gas

$$H = U + pV = C_v^0 T + RT, \tag{6.11}$$

where C_v^0 is the heat capacity of one mole of gas.

According to Equation (3.23), then,

$$C_p^0 = \left(\frac{\partial H}{\partial T}\right)_p = C_v^0 + R, \tag{6.12}$$

where C_p^0 likewise corresponds to one mole of gas. Thus

$$C_p^0 - C_v^0 = R, \tag{6.13}$$

so that the aggregate of functions on the right of Equation (6.10) should equal R. By substituting Equation (5.1) into both Equation (6.1) and Equation (6.2) (for one mole of gas), we obtain

$$\alpha = \frac{1}{T}, \tag{6.14}$$

$$\beta = \frac{1}{p}. \tag{6.15}$$

Thus, for the combination on the right of Equation (6.10) we have

$$\frac{VT\alpha^2}{\beta} = \frac{VT\left(\frac{1}{T}\right)^2}{\frac{1}{p}} = \frac{pV}{T} = R, \tag{6.16}$$

which is the predicted result. This confirms Equation (6.10) in the special case of the ideal gas.

2. Energetics of the Free Expansion of a Gas

It is of interest to determine the temperature change experienced by a gas when it undergoes expansion in such a manner that its internal energy remains constant. In effect, we are interested in the coefficient $\left[\frac{\partial T}{\partial V}\right]_U$. It is not *necessary* that such an expansion take place in an *isolated* system, but an expansion in an isolated system *will conform* to the requirement that U remain unchanged. For example, let us deal with the system of Figure 6.1. In this system a gas is confined initially by a membrane to the left side of a container possessing rigid adiabatic walls which are impermeable to matter.

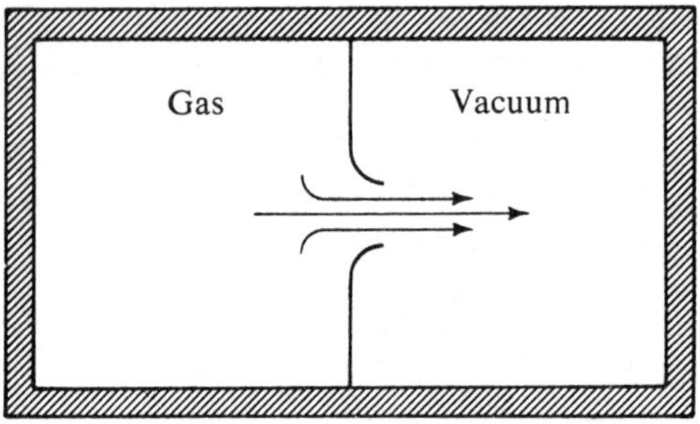

FIGURE 6.1. *Expansion of a gas in an isolated system. The cross-hatched walls are rigid and adiabatic, and the gas expands through the hole in the punctured membrane as shown by the arrows indicating the flow paths.*

As a result, the system is isolated. The expansion may be induced by puncturing the membrane so that the gas moves into the vacuum on the right side of the vessel. The change of temperature for this process would be obtained by integrating the coefficient $\left[\frac{\partial T}{\partial V}\right]_U$ between volume on the left of the container and the final volume—the total volume of the containing vessel. To derive a useful expression for $\left[\frac{\partial T}{\partial V}\right]_U$ we proceed as follows. We employ Equation (2.36)

$$\left[\frac{\partial T}{\partial V}\right]_U = -\left[\frac{\partial T}{\partial U}\right]_V \left[\frac{\partial U}{\partial V}\right]_T, \tag{6.17}$$

$$= \frac{-\left[\frac{\partial U}{\partial V}\right]_T}{C_v}, \tag{6.18}$$

where the definition of C_v given in Equation (3.19) has been used. Substitution of Equation (6.5) into the numerator on the right of Equation (6.18) gives

$$\left[\frac{\partial T}{\partial V}\right]_U = \frac{p - T\left[\frac{\partial S}{\partial V}\right]_T}{C_v}, \tag{6.19}$$

and substituting Equation (4.87) into the numerator of Equation (6.19) then yields

$$\left[\frac{\partial T}{\partial V}\right]_U = \frac{p - T\left[\frac{\partial p}{\partial T}\right]_V}{C_v}. \tag{6.20}$$

This is the expression we have been seeking. If Equation (6.20) is applied to the special case of one mole of an ideal gas, we obtain, using Equation (5.1),

$$\left[\frac{\partial p}{\partial T}\right]_V = \frac{R}{V}. \tag{6.21}$$

By substituting this into Equation (6.20) we get

$$\left[\frac{\partial T}{\partial V}\right]_U = \frac{p - \dfrac{RT}{V}}{C_v} = 0, \tag{6.22}$$

since $\dfrac{RT}{V} = p$, according to Equation (5.1). Thus, the free expansion of an isolated ideal gas occasions no change in temperature. This result is the expected one since we have seen that the internal energy of an ideal gas depends only on temperature, and the reverse is also true.

3. The Joule-Thomson Coefficient

An interesting quantity whose magnitude has been used as an indicator of the degree of departure of a gas from the ideal state is the Joule-Thomson coefficient. This quantity is very similar to $\left[\frac{\partial T}{\partial V}\right]_U$, being $\left[\frac{\partial T}{\partial p}\right]_H$. One may derive an expression for it in the following manner.

Equation (4.79) is differentiated (in the manner of Equation (2.22)) with respect to p, holding T constant. The result is

$$\left[\frac{\partial H}{\partial p}\right]_T = T\left[\frac{\partial S}{\partial p}\right]_T + V \tag{6.23}$$

or

$$\left[\frac{\partial S}{\partial p}\right]_T = \frac{1}{T}\left\{\left[\frac{\partial H}{\partial p}\right]_T - V\right\} = -\left[\frac{\partial V}{\partial T}\right]_p,$$

where Equation (4.92) has been employed to produce the last equation.

We employ Equation (2.36) and get

$$\left[\frac{\partial T}{\partial p}\right]_H = -\left[\frac{\partial T}{\partial H}\right]_p\left[\frac{\partial H}{\partial p}\right]_T = \frac{-\left[\dfrac{\partial H}{\partial p}\right]_T}{\left[\dfrac{\partial H}{\partial T}\right]_p}. \tag{6.24}$$

Now according to Equation (3.23), the denominator in Equation (6.25) is C_p. Furthermore, the numerator is given by the right-hand equation of the second line in Equation (6.23). By making these substitutions in Equation (6.24) we obtain

$$\left[\frac{\partial T}{\partial p}\right]_H = \frac{\left\{T\left[\frac{\partial V}{\partial T}\right]_p - V\right\}}{C_p}. \tag{6.25}$$

This is the expression we have been seeking.

Once again it is instructive to introduce the ideal gas law (5.1). From this we see that for one mole

$$T\left[\frac{\partial V}{\partial T}\right]_p = \frac{RT}{p} = V, \tag{6.26}$$

so that the numerator in Equation (6.25), and consequently $\left[\frac{\partial T}{\partial p}\right]_H$, is zero for an ideal gas.

The Joule-Thomson coefficient has achieved popularity because it is a conveniently measurable entity. For actual gases it is negative at high temperatures, becoming positive at lower temperatures. The intermediate temperature at which it is zero is known as the inversion temperature.

The measurement is performed using an apparatus which is illustrated schematically in Figure 6.2. It consists of two compartments, each bounded by a piston and separated by a porous plug. The pistons and the walls of the vessel are constructed of thermally insulating material. Thus, whatever process takes place within the vessel will do so adiabatically. The gas whose Joule-Thomson coefficient is to be measured is placed in each compartment, and the pistons are moved in the directions indicated by the arrows so as to maintain the pressures in the left and right compartments at p_1 and p_2, respectively. The directions of movement indicated by the arrows are consistent with $p_1 > p_2$.

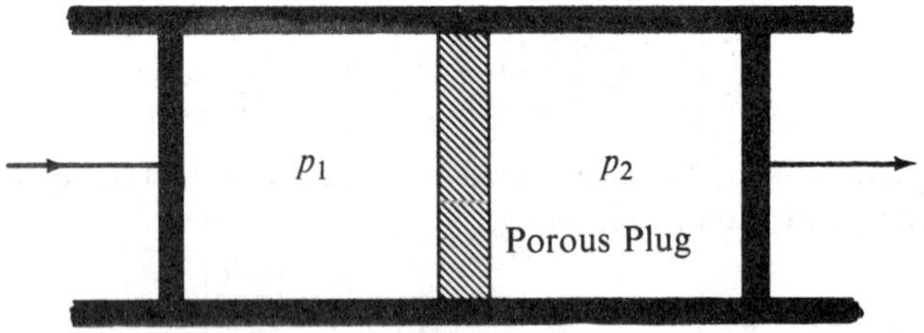

FIGURE 6.2. *Apparatus for performing the Joule-Thomson experiment. The pistons are moved slowly in the directions indicated by the arrows so as to maintain the pressures p_1 and p_2 as the gas transpires through the porous plug. All walls, including the pistons, are composed of adiabatic materials, so that the gas is thermally insulated from its surroundings.*

As a result of the pressure differential, gas will transpire through the porous plug from left to right. If the pistons are constantly monitored, a continuous process may be set up in which the pressures p_1 and p_2 are maintained constant. The purpose of the porous plug is to permit the process to proceed slowly enough so that the gases in the two compartments remain effectively at equilibrium, and definable pressures are established (see the discussion following Equation (3.21) or that preceding Equation (4.26) in Chapter 4, Section 7).

Suppose the pressures are well defined, and the volume of a mole of gas on the left side is denoted by V_1 while the same volume on the right is V_2. Thus, when one mole of gas has been displaced from the left side, the piston on that side will have performed the volume work p_1V_1, while the piston on the right will have had the work p_2V_2 performed *on* it. Thus, the net work performed *by* the gas will be

$$\Delta w = p_2V_2 - p_1V_1. \tag{6.27}$$

At the same time, one mole of gas having the internal energy U_1 on the right will have been displaced to the left were its internal energy will now be U_2. Thus, the change in internal energy for this process is

$$\Delta U = U_2 - U_1. \tag{6.28}$$

Since the process is adiabatic $\Delta q = 0$, and according to Equation (3.18),

$$\Delta U = U_2 - U_1 = -\Delta w = p_1V_1 - p_2V_2, \tag{6.29}$$

where Equations (6.27) and (5.28) have been used. This equation may be rearranged to yield

$$U_2 + p_2V_2 = U_1 + p_1V_1 \tag{6.30}$$

or

$$H_2 = H_1,$$

where Equation (3.20) has been employed. Equation (6.30) demonstrates that the transpiration process is one in which the enthalpy is conserved.

It is clear that the gases on both sides of the porous plug having different pressures and molar volumes may have different temperatures T_1 and T_2, respectively. If these are measured, then it should be clear that

$$\left[\frac{\partial T}{\partial p}\right]_H = \lim_{p_2 \to p_1}\left\{\frac{T_2 - T_1}{p_2 - p_1}\right\}, \tag{6.31}$$

for the process in question. Thus, the Joule-Thomson coefficient may be measured.

The Joule-Thomson coefficient is very useful for calibrating a gas thermometer in terms of the absolute temperature scale. For a detailed discussion of this process, the reader is referred to Pippard's book.[7]

[7] A. B. Pippard, *Elements of Classical Thermodynamics* (New York: Cambridge University Press, 1960), pp. 87–93.

4. Relation of Enthalpy and Entropy to Heat Capacity

The measurement of heat capacity as a function of temperature permits one to determine both the enthalpy and entropy of a substance. In the case of enthalpy, this is accomplished through the integration of Equation (3.23) leading to the result

$$H(T,p_0) = H(T_0,p_0) + \int_{T_0}^{T} C_p(T,p_0)\, dT. \tag{6.32}$$

In Equation (6.32) the variables T_0, p_0 characterize an arbitrary reference state and $C_p(T,p_0)$ is the heat capacity, always measured at pressure p_0.

Equation (6.32) must be modified when phase transitions such as ordinary melting occur, for C_p is not defined at such isolated singular points. On the other hand, according to Chapter 3, Section 6, the heat absorbed at constant pressure is also the change in H. Thus, for a phase transition, like melting, the heat of fusion λ_f measures the increment in H. If the range (T_0,T) includes the melting point T_f, then Equation (6.32) must be amended to read

$$H(T,p_0) = H(T_0,p_0) + \int_{T_0}^{T_f} C_p(T,p_0)\, dT + \lambda_f + \int_{T_f}^{T} C_p(T,p_0)\, dT. \tag{6.33}$$

After having measured the enthalpy, the internal energy U may be computed by subtracting p_0V from the value of H so obtained.

Knowledge of the heat capacity may also be used in the determination of the entropy. This time we integrate Equation (6.7) to obtain

$$S(T,p_0) = S(T_0,p_0) + \int_{T_0}^{T_f} \frac{C_p(T,p_0)}{T}\, dT + \frac{\lambda_f}{T} + \int_{T_f}^{T} \frac{C_p(T,p_0)}{T}\, dT. \tag{6.34}$$

We shall have more to say about these relations when we discuss the *third law*.

5. Magnetic Substances

The process of cooling below 1°K is made possible by the phenomenon of *adiabatic demagnetization* which was suggested almost simultaneously by Giauque and Debye in 1926. We shall undertake the quantitative discussion of this phenomenon in the chapter on the *third law*. The treatment can only proceed through a knowledge of the thermodynamics of paramagnetic substances, and a few thermodynamic formulas concerning such substances will be derived in the present section.

A paramagnetic substance may be involved in magnetic as well as volume work, and so Equation (4.74) must be modified by the addition of the work term which appears in Equation (1.5). The result is

$$dU = T\, dS - p\, dV + \mathscr{H}\, d\mathscr{M}. \tag{6.35}$$

It is convenient to focus attention on the intensive variables T, p, and $\mathscr{H}$ as independent variables, and so we synthesize a new function of state P which will generate a Legendre transformation (see Chapter 4, Section 13) passing from S, V, and $\mathscr{M}$ to T, p, and $\mathscr{H}$. This new function has the form

$$P = U + pV - \mathscr{H}\mathscr{M} - TS. \tag{6.36}$$

The total differential of P is

$$\begin{aligned} dP &= dU + p\,dV + V\,dp - \mathscr{H}\,d\mathscr{M} - \mathscr{M}\,d\mathscr{H} - T\,dS - S\,dT \\ &= V\,dp - \mathscr{M}\,d\mathscr{H} - S\,dT, \end{aligned} \tag{6.37}$$

where the last form is obtained by substituting the right side of Equation (6.35) for dU.

Since P is a function of state, it is now possible to derive Maxwell reciprocity relations through the use of Equation (2.6). Thus,

$$\left[\frac{\partial V}{\partial \mathscr{H}}\right]_{T,p} = -\left[\frac{\partial \mathscr{M}}{\partial p}\right]_{T,\mathscr{H}}, \tag{6.38}$$

$$\left[\frac{\partial V}{\partial T}\right]_{\mathscr{H},p} = -\left[\frac{\partial S}{\partial p}\right]_{\mathscr{H},T}. \tag{6.39}$$

$$\left[\frac{\partial \mathscr{M}}{\partial T}\right]_{\mathscr{H},p} = \left[\frac{\partial S}{\partial \mathscr{H}}\right]_{T,p}. \tag{6.40}$$

The last of these is the one of importance for the study of cooling by adiabatic demagnetization (see Chapter 15).

The relations derived in this chapter furnish excellent examples of the kinds of interconnections between variables which thermodynamics can provide. The role of the exact differential, and therefore of the function of state, in facilitating these derivations cannot be overemphasized. Attention is therefore drawn, once again, to the importance of having a large catalog of state functions.

VII

Internal Equilibrium and the Extremal Properties of the Entropy

1. Extremal Condition on the Entropy

It was emphasized in Section 5 of Chapter 4 that the inequality (4.16) referred to a process which was the *inverse* of one which might have occurred in an isolated system. The process envisioned (in which $\Delta S < 0$) is a *virtual* variation (Chapter 4, Section 6) in which the system is driven away from a state of equilibrium attained in the isolated system. A great deal of emphasis was placed on the "inverse" character of the process in order to make clear the fact that the isolation of the system must be violated to conduct the inverse process. This is necessary so that an additional constraint (χ_1 of Chapter 1, Section 8) can be introduced in order to drive the system away from its initial equilibrium state. In driving the system, the new constraint performs work, and therefore the isolation is destroyed.

In Chapter 4, Section 6, the point is made that the process is one in which the initial set of constraints χ_0 (Chapter 1, Section 8), subject to which equilibrium is attained, is retained. This only means that the virtual variation is caused by the addition of rather than the removal of constraints. Since $\Delta S < 0$ for all variations (beginning with the equilibrium state) which meet the requirements of inverseness and retention of constraints, it is clear that *along such paths* S is a maximum at the initial equilibrium state. This fact has been signalized in Chapter 4, Section 6, by the differential condition $\delta S = 0$, where the infinitesimal variation involved is one which begins in equilibrium and follows one of the acceptable inverse paths.

Now we have used the phrase "begins in equilibrium," and therefore as the discussion of Chapter 4, Section 6 indicates, we are well on our way to committing the common sin which adds so much confusion to thermodynamics. The use of such terminology implies that the *initial* state is one of equilibrium, but that the states through which the constraint drives the virtual variation are not.

Actually, as it is emphasized in Chapter 4, entropy is a function of state and cannot even be defined or measured (at least not by the methods of thermodynamics) in a nonequilibrium (nonthermodynamic) state. Therefore, if a displacement to a nonequilibrium state occurred, neither ΔS nor δS could be defined and the inequality (4.16) would have no meaning. What is really meant is the following. The new constraint $\mathcal{X}_1$, which is introduced in order to drive the system through the variation, drives it over a perfectly acceptable path in *state space*, that is, through a sequence of equilibrium states. In Chapter 1 it was demonstrated that constraints and independent variables are in one-to-one correspondence so that the application of the new constraint increases the number of independent state variables. As a result, the system is displaced from a state of equilibrium in which it possessed a certain number, say n independent variables, to a new equilibrium state in which there are $n + 1$ variables. This is the significance of the virtual variation. It means displacement from an equilibrium state controlled by a definite number of variables to one controlled by a larger number. *Throughout the displacement the system remains in equilibrium.*

2. A One-Component, Two-Phase System

The extremal properties of the entropy will now be used to study certain aspects of two phases which are in equilibrium with each other. Consider an isolated thermodynamic system consisting of two phases and one component. A liquid in contact with its vapor enclosed in an impermeable rigid adiabatic container furnishes one example of such a system. Any spontaneous process occurring within this system must occur subject to the constraints of isolation. These require:

(1) the internal energy U to remain fixed;
(2) the volume V to remain fixed so that no work is exchanged with the environment;
(3) the number of moles n of whatever species composes the system to remain fixed.

The requirement of adiabaticity need not be stated because if no work is performed, and U remains constant, Equation (3.18) requires $Dq = 0$. These are the constraints $\mathcal{X}_0$, subject to which the system comes to equilibrium; and if we are to have an inverse process (virtual variation) so that $\delta S = 0$ can be used, these constraints must be retained. We symbolize this by writing

$$(\delta S)_{U,V,n} = 0, \tag{7.1}$$

where the subscripts U, V, and n indicate that these variables are held constant.

Denote the two phases in the system as phases 1 and 2, respectively. Although the entire system is *closed* in the sense that it cannot exchange matter with its environment, each phase is *open* since matter may be transferred from one phase to another. It will be recalled that Equation (3.18) and (4.74) are limited to closed systems. If phase 1, for example, were closed, Equation (4.74) could be applied in the form

$$dU^{(1)} = T^{(1)}\,dS^{(1)} - p^{(1)}\,dV^{(1)}, \tag{7.2}$$

where the superscripts indicate that the various quantities refer to phase 1. Since the phase is actually an open system, its state depends not only on the two variables $S^{(1)}$ and $V^{(1)}$, but also upon $n^{(1)}$, the number of moles in phase 1. We must therefore write†

$$dU^{(1)} = T^{(1)}\,dS^{(1)} - p^{(1)}\,dV^{(1)} + \mu^{(1)}\,dn^{(1)}, \tag{7.3}$$

where

$$\mu^{(1)} = \left[\frac{\partial U^{(1)}}{\partial n^{(1)}}\right]_{S^{(1)},V^{(1)}}, \tag{7.4}$$

and as in Equations (4.75) and (4.76)

$$T^{(1)} = \left[\frac{\partial U^{(1)}}{\partial S^{(1)}}\right]_{V^{(1)},n^{(1)}}, \tag{7.5}$$

$$p^{(1)} = -\left[\frac{\partial U^{(1)}}{\partial V^{(1)}}\right]_{S^{(1)},n^{(1)}}. \tag{7.6}$$

For phase 2 a similar relation holds,

$$dU^{(2)} = T^{(2)}\,dS^{(2)} - p^{(2)}\,dV^{(2)} + \mu^{(2)}\,dn^{(2)}, \tag{7.7}$$

where $T^{(2)}$, $p^{(2)}$, and $\mu^{(2)}$ generate relations analogous to Equations (7.4), (7.5), and (7.6).

Now for the total system we have

$$U = U^{(1)} + U^{(2)}, \tag{7.8}$$

$$S = S^{(1)} + S^{(2)}, \tag{7.9}$$

$$V = V^{(1)} + V^{(2)}, \tag{7.10}$$

$$n = n^{(1)} + n^{(2)}. \tag{7.11}$$

† Some authors refer to the term $\mu^{(1)}\,dn^{(1)}$ as "chemical" work because of the manner of its appearance in Equation (7.3). This is loose and misleading because there is no *direct* mechanical method for measuring $\mu^{(1)}\,dn^{(1)}$. As a result, it is best to regard the new term simply as a correction required to account for the openness of the system.

It is possible to solve both Equations (7.3) and (7.7) for $dS^{(1)}$ and $dS^{(2)}$, respectively. Thus,

$$dS^{(1)} = \frac{1}{T^{(1)}}\, dU^{(1)} + \frac{p^{(1)}}{T^{(1)}}\, dV^{(1)} - \frac{\mu^{(1)}}{T^{(1)}}\, dn^{(1)}, \tag{7.12}$$

$$dS^{(2)} = \frac{1}{T^{(2)}}\, dU^{(2)} + \frac{p^{(2)}}{T^{(2)}}\, dV^{(2)} - \frac{\mu^{(2)}}{T^{(2)}}\, dn^{(2)}. \tag{7.13}$$

Therefore,

$$\begin{aligned} dS &= dS^{(1)} + dS^{(2)} \\ &= \frac{1}{T^{(1)}}\, dU^{(1)} + \frac{1}{T^{(2)}}\, dU^{(2)} + \frac{p^{(1)}}{T^{(1)}}\, dV^{(1)} + \frac{p^{(2)}}{T^{(2)}}\, dV^{(2)} \\ &\quad - \frac{\mu^{(1)}}{T^{(1)}}\, dn^{(1)} - \frac{\mu^{(2)}}{T^{(2)}}\, dn^{(2)}. \end{aligned} \tag{7.14}$$

If we now restrict the differential in Equation (7.14) to the constrained path required by Equation (7.1), we have

$$\delta U = \delta U^{(1)} + \delta U^{(2)} = 0, \tag{7.15}$$

$$\delta V = \delta V^{(1)} + \delta V^{(2)} = 0, \tag{7.16}$$

$$\delta n = \delta n^{(1)} + \delta n^{(2)} = 0, \tag{7.17}$$

since U, V, and n are held constant over such paths. Notice that we now use δ to denote the virtual variation. Upon introducing Equations (7.15), (7.16), and (7.17) into Equation (7.14) and the result into Equation (7.1), we have

$$\begin{aligned} (\delta S)_{U,V,n} &= \left[\frac{1}{T^{(1)}} - \frac{1}{T^{(2)}}\right] \delta U^{(1)} + \left[\frac{p^{(1)}}{T^{(1)}} - \frac{p^{(2)}}{T^{(2)}}\right] \delta V^{(1)} \\ &\quad - \left[\frac{\mu^{(1)}}{T^{(1)}} - \frac{\mu^{(2)}}{T^{(2)}}\right] \delta n^{(1)} = 0. \end{aligned} \tag{7.18}$$

Since $\delta U^{(1)}$, $\delta V^{(1)}$, and $\delta n^{(1)}$ are independent variations, Equation (7.18) can only hold if their coefficients vanish identically (see Chapter 2, Section 6). Therefore, we arrive at the following results:

$$\frac{1}{T^{(1)}} = \frac{1}{T^{(2)}}, \tag{7.19}$$

$$\frac{p^{(1)}}{T^{(1)}} = \frac{p^{(2)}}{T^{(2)}}, \tag{7.20}$$

$$\frac{\mu^{(1)}}{T^{(1)}} = \frac{\mu^{(2)}}{T^{(2)}}. \tag{7.21}$$

Together, these conditions require

$$T^{(1)} = T^{(2)}, \tag{7.22}$$

$$p^{(1)} = p^{(2)}, \tag{7.23}$$

$$\mu^{(1)} = \mu^{(2)}. \tag{7.24}$$

3. "Feature" of Equilibrium

Very often the condition (7.1) is referred to as a "criterion" of equilibrium. If we retrace the steps leading to Equations (7.22), (7.23), and (7.24), it is evident that we began with a composite two-phase *isolated* system *known to have reached equilibrium*. Therefore, we had no interest in testing the system to determine whether or not it was in equilibrium. Stated in another way, *we had no immediate use for a "criterion" of equilibrium which could be employed in such a test.*

As the discussion of Chapter 4, Section 6 indicates, the new constraint χ_1, which causes the virtual variation, need not be specified. Correspondingly, the detailed character of the new state to which the system is displaced by the variation need not be specified. This is made clear in the expanded form of Equation (7.1) which appears in Equation (7.18). From this equation the valuable relations (7.22), (7.23), and (7.24) are derived from an analysis of the coefficients. These coefficients refer to the properties of the *initial* state (provided that the final state is only infinitesimally removed from the initial one) and make no demands on the *final* state to which the system is displaced. They provide information about the internal condition of the system in the initial state. Thus, we learn that when the composite isolated system achieves equilibrium, the temperature must be the same in both phases, and that the same must be true of the pressure and the quantity μ. Use was therefore made of the "feature" of equilibrium embodied in Equation (7.1) to derive information about the system in equilibrium subject to the initial constraints χ_0.

In one possible example of how the virtual variation might be conducted, an impermeable but diathermic partition might be inserted between the two phases and moved so as to compress one of them (say the liquid phase if the system consists of a liquid and its vapor). It is clear that $p^{(1)}$ will then no longer equal $p^{(2)}$. As a matter of fact, it can be shown that such compression will cause $\mu^{(1)}$ to cease to equal $\mu^{(2)}$, but this point will not be elaborated here. Since the partition is diathermic (conducts heat), T_1 will continue to equal T_2.

If the compression of the liquid phase (and consequent expansion of the vapor phase) is carried out so that U, V, and n remain constant, then the entropy of the system must decrease for a finite process, and remain invariant (according to Equation (7.1)) for an infinitesimal process. The confining partition is the new constraint χ_1. The new state (under compression) to which the system is displaced is still *an equilibrium* state, but it is one in which the single variable V is not sufficient for the description of the state. Instead, we must use V_1 and V_2 separately. Thus, the new equilibrium state has more independent variables than the initial one.

4. Internal Potentials

T, p, and μ in Equations (7.22), (7.23), and (7.24) may be called *internal potentials* because they determine when and in what direction certain extensive quantities will flow within the system as it moves to a new state when a constraint is lifted. These statements will be proved later when the question of "stability" is considered. In this section our comments will be limited to qualitative discussion.

The condition (7.24) is a direct consequence of the fact that matter could be transferred between phases 1 and 2 of the system in question. This "openness" of the two phases is responsible for the appearance of both $dn^{(1)}$ and $dn^{(2)}$ on the right-hand sides of Equations (7.12) and (7.13). Had there been a restriction preventing the transfer of matter, these terms, with coefficients $\mu^{(1)}/T^{(1)}$ and $\mu^{(2)}/T^{(2)}$ would not have appeared; and no condition such as Equation (7.24) would have emerged from the resulting development. Thus, the system may achieve a state of equilibrium in which $\mu^{(1)} \neq \mu^{(2)}$, provided that matter cannot be transferred between phases 1 and 2.

As we shall show, in Chapter 14, when the constraint forbidding the transfer of matter is lifted, matter will flow from the region of higher μ to that of lower μ. This means that if $\mu^{(2)} > \mu^{(1)}$, matter will flow from phase 2 to phase 1 until a new equilibrium is established in which $\mu^{(1)} = \mu^{(2)}$. Therefore, μ plays the role of a potential which determines the direction of flow of matter. It has been given the name *chemical potential.*

In a similar manner, it may be shown that when $T^{(2)} > T^{(1)}$ entropy will flow from phase 2 to phase 1. Temperature, therefore, acts as an internal potential governing the flow of entropy. Furthermore, an equilibrium in which $T^{(2)} \neq T^{(1)}$ can only be achieved when a constraint, which prevents the transfer of entropy between phases 1 and 2, is applied.

In general, equilibrium in which internal potentials are not uniform can only be achieved when constraints exist, forbidding the transfer of conjugate quantities.

The negative of pressure, $-p$, plays the role of a potential for the flow of volume. Figure (7.1) will be helpful in understanding what is meant by the transfer of volume. Initially, the system in the figure is divided by a partition into two compartments with volumes $V^{(2)}$ and $V^{(1)}$, respectively; the partition

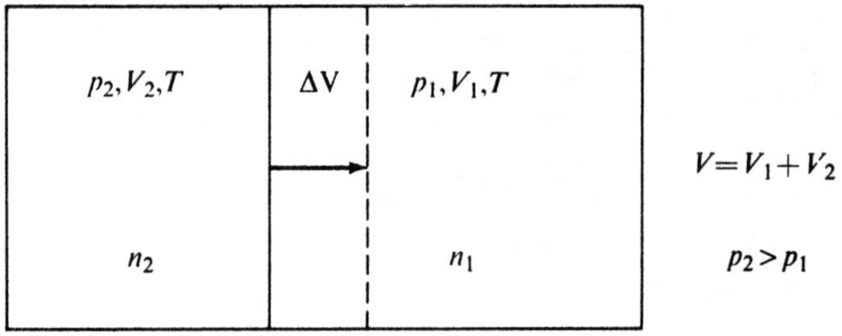

FIGURE 7.1. *Diagram illustrating the transfer of volume between two regions having different pressures.*

is locked in place so that $p^{(2)} > p^{(1)}$. When the lock is released, the pressures will equalize and the partition will move to the right to the dashed line so that the additional volume ΔV appears on its left in the compartment where the pressure is $p^{(2)}$. At the same time, the volume ΔV will be abstracted from the side where the pressure is $p^{(1)}$. Thus, the volume ΔV will have been transferred from side 1 to side 2. Since $p^{(2)} > p^{(1)}$ or $-p^{(1)} > -p^{(2)}$, ΔV moves from the region of high $-p$ to that of low $-p$. Thus, $-p$ acts as an internal potential for the transfer of volume. Once again, the system can be in equilibrium, with $p^{(1)} \neq p^{(2)}$, only if there is a constraint against the transfer of the conjugate quantity, volume.

VIII

Thermodynamic Potentials

1. Concept of the Thermodynamic Potential

In our development, Equation (7.1) has emerged as a consequence of the inequality (4.16). It represents the condition that S shall have an extremum in the state of equilibrium determined by the constraints χ_0. Since S is maximized in this particular state, it behaves like the *mechanical potential* behaves in respect to *mechanical* equilibrium, except that potential energy is *minimized* rather than *maximized.* As a result, the entropy is sometimes referred to as a *thermodynamic potential.* In view of the useful relation (7.22), (7.23), and (7.24), all of which may be traced to Equation (7.1), it is clear that the thermodynamic potential is a valuable concept. It is therefore natural to inquire whether thermodynamic potentials other than entropy exist. The answer is affirmative, and the present chapter will be devoted to this problem.

With the possible exception of entropy itself, no other concept in thermodynamics has occasioned so much confusion as the thermodynamic potential. The most distinguished, as well as less distinguished authors have propagated what is a manifestly false doctrine. The confusion is compounded by the fact that it is possible to derive the correct relations with the incorrect philosophy. As we shall see, part of the reason for this lies in the circumstance that the detailed character of the new equilibrium state, to which a system is displaced by a virtual variation, need not be specified. Another part of the reason is the high degree of symmetry found among thermodynamic relations so that new formulas are frequently derivable by analogy with old ones.

Unfortunately, however, it is always possible to make mistakes through such reasoning-by-analogy, and some important errors have been committed. Since much of the value of thermodynamics lies in the almost absolute character of the information which it provides, one can hardly afford unnecessary sacrifices of rigor. One of the uses to which thermodynamic potentials are put is the derivation of formulas such as Equation (7.24), which specifies the condition under which matter will exhibit no tendency to flow between two phases in contact. If one only cared about deriving Equation (7.24), it would hardly be necessary to set in motion the rather sophisticated machinery of thermodynamic potentials. We shall demonstrate this in the next section, but before doing so it is important to assure the reader that there are other uses for such potentials. They make possible the discussion of "stability," and the separate treatment of *stable* or *unstable* equilibriums (see Chapter 14); whereas the elementary method presented in the next section permits us to derive relations like Equation (7.24) which are valid for *either* type of equilibrium. Furthermore, the method of thermodynamic potentials provides the theory with an essential unity, and also makes possible the study of the "direction of change."

2. An Elementary Method

The material in this chapter will draw very heavily on the concepts developed in Sections 5 through 8 of Chapter 1 and Sections 6 through 8 of Chapter 4. The reader may wish to review these sections to make certain that he has firmly fixed in his mind such ideas as the macroscopic state space, the quasistatic process, the correspondence between constraints, variables, and work, and finally the meaning of the virtual variation. In this section we shall concentrate especially on state space and quasistatic processes. We wish to derive condition (7.24) for the two-phase, one-component system discussed in Chapter 7. As in Chapter 7, we will require that the system be closed so that it cannot exchange matter with its surroundings. Our objective will be to derive Equation (7.24) without employing thermodynamic potentials.

The differential of the internal energy for the entire system may be expressed, in accordance with Equation (3.18), as

$$dU = Dq_{\text{rev}} - Dw_{\text{rev}}, \tag{8.1}$$

where Dq_{rev}, the heat absorbed, and Dw_{rev}, the work performed, are the values realized when the infinitesimal change is conducted reversibly. This restriction means that a path in state space is involved. (As usual, dU has the same value whether the change is conducted reversibly or not.)

Now since phases 1 and 2 may exchange matter, they are both *open*, and the differentials of internal energy take the forms

$$dU^{(1)} = Dq_{\text{rev}}^{(1)} - Dw_{\text{rev}}^{(1)} + \mu^{(1)}\, dn^{(1)}, \tag{8.2}$$

$$dU^{(2)} = Dq_{\text{rev}}^{(2)} - Dw_{\text{rev}}^{(2)} + \mu^{(2)}\, dn^{(2)}, \tag{8.3}$$

where $\mu^{(1)}$ and $\mu^{(2)}$ have the same meaning as in Chapter 7. Equation (8.2) is identical with Equation (7.3) because $T^{(1)}\, dS^{(1)}$ measures the heat absorbed reversibly and $p^{(1)}\,dV^{(1)}$ is the reversible work, provided that $p^{(1)}$ is the equilibrium pressure. The same is true of Equation (8.3) and the counterpart of Equation (7.3) for phase 2.

We have deliberately chosen the forms (8.2) and (8.3) rather than (7.3) so that our argument will remain valid if the system is generalized so as to perform other work besides the *volume* work $p^{(1)}\,dV^{(1)}$ and $p^{(2)}\,dV^{(2)}$. As we saw in Chapter 1, Section 8, the system may be organized to perform *new* kinds of work through the addition of *new* constraints. Each new constraint adds another variable, and therefore $Dw_{\text{rev}}^{(1)}$ and $Dw_{\text{rev}}^{(2)}$ will contain other differentials besides $dV^{(1)}$ and $dV^{(2)}$. Nevertheless, Equations (8.2) and (8.3) remain valid.

The reason for the limitation of the heats and works to reversible processes is clear. We wish to express $dU^{(1)}$ and $dU^{(2)}$ as complete differentials in terms of their independent variables, and terms like $T^{(1)}\,dS^{(1)}$ and $p^{(1)}\,dV^{(1)}$ are essential to this purpose. Once the form (7.3) has been achieved, it becomes unnecessary to refer explicitly to heat or work. $dU^{(1)}$ is merely an exact differential expressed in terms of the given set of independent variables. In Equation (8.2), for example, if the heat and work terms were not those corresponding to a reversible change, the last term could not be $\mu^{(1)}\,dn^{(1)}$ but instead it would be some other vague contribution.

Since Equations (8.2) and (8.3) refer to variations conducted over paths in state space, that is, variations through a sequence of equilibrium states, all the coefficients T, p, μ, et cetera, of the differentials on the right of these equations then refer to equilibrium states; *and any relation between them is an equilibrium relation.* This is the crucial point in the present development.

Now

$$Dq_{\text{rev}} = Dq_{\text{rev}}^{(1)} + Dq_{\text{rev}}^{(2)}, \tag{8.4}$$

$$Dw_{\text{rev}} = Dw_{\text{rev}}^{(1)} + Dw_{\text{rev}}^{(2)}, \tag{8.5}$$

$$dU = dU^{(1)} + dU^{(2)}. \tag{8.6}$$

By adding Equation (8.2) to Equation (8.3) and applying Equations (8.4), (8.5), and (8.6) to the result we get

$$dU = Dq_{\text{rev}} - Dw_{\text{rev}} + \mu^{(1)}\,dn^{(1)} + \mu^{(2)}\,dn^{(2)}; \tag{8.7}$$

and comparison of this with Equation (8.1) gives

$$\mu^{(1)}\,dn^{(1)} + \mu^{(2)}\,dn^{(2)} = (\mu^{(1)} - \mu^{(2)})\,dn^{(1)} = 0, \tag{8.8}$$

where Equation (7.17) has been used since the system is closed. The solution of Equation (8.8) is

$$\mu^{(1)} = \mu^{(2)}. \tag{8.9}$$

Now we have asserted that the crucial point of the development is the fact that any relation among the differential coefficients of Equations (8.2) and (8.3) is an *equilibrium* relation (since (8.2) and (8.3) refer to paths in state space), and therefore Equation (8.9) is an equilibrium relation. It is identical with Equation (7.24). It has not been necessary to introduce thermodynamic potentials to prove this point.

3. Sign of the Work Performed by the Variational Constraint

In this section we return to the subject of thermodynamic potentials. If a system is divided into subsystems (phases, for example), then it is conceivable that one might apply constraints in such a manner that subsystems which communicate with one another, in the sense that matter may be exchanged between them, can be in mutual equilibrium even though intensive variables like temperature and pressure may have different values in each subsystem. In the case of pressure, there is the familiar example of osmosis (discussed in Chapter 1, Section 8) in which equilibrium is established between two phases at different pressures which exchange matter through a semipermeable membrane. Unfortunately, there is no known method by means of which two phases with differing temperatures may be constrained into *equilibrium* while they are permitted to exchange matter. (The cases of thermo-osmosis[8] or thermal diffusion[8] are nonequilibrium phenomena in which there are continuous flows of heat between the two phases at different temperatures.) As a result, there is not much point to developing thermodynamic potentials for systems with nonuniform distributions of temperature, and in this book only systems with uniform temperature will be considered.

Let us recall the discussion, in Section 8 of Chapter 1, of the sets of constraints $\mathcal{X}_0$ and $\mathcal{X}_1$. We recall that $\mathcal{X}_0$ are the constraints subject to which the system came to equilibrium (they might be called the *initial* constraints), while $\mathcal{X}_1$ is the constraint causing the variation (it might be called the *variational* constraint). During a variation, the work performed *on* the system by both sets of constraints is $-(\Delta w_0 + \Delta w_1)$. Now the virtual variation is a reversible process (see Chapter 4, Section 6) so that the heat absorbed by the system during the variation is measured by $\int T\,dS$, where the integral is used because a finite process is involved. According to Equation (3.18) then, the change in internal energy (for a closed system) during the variation is

$$\Delta U = \int T\,dS - \Delta w_{0_{\text{rev}}} - \Delta w_{1_{\text{rev}}}. \tag{8.10}$$

[8] S. R. deGroot and P. Mazur, *Nonequilibrium Thermodynamics* (Amsterdam: North Holland Publishing Company, 1962), pp. 273–284, 431–432.

Now our intuition tells us that since $\mathcal{X}_1$ is *causing* a displacement of the system from its initial state of equilibrium† it must be *performing* work *on* the system so that

$$-\Delta w_{1\text{rev}} > 0. \tag{8.11}$$

Our intuition is right, and Equation (8.11) can be shown to be true. However, we cannot rely on intuition alone but must *prove* the validity of (8.11) through an application of the first and second laws of thermodynamics. These laws are presented in a form such that the fundamental inequality is (4.2) and involves the *entropy* rather than the *work*. Thus, adherence to the strict principles of logic demands that the fundamental, almost self-evident, embodiment of Equation (8.11) be derived from the inequality involving entropy. This involves some manipulation which we shall perform in the next section. For the moment, however, let us dwell on some of the consequences of (8.11).

Equation (8.10) may be rearranged as follows:

$$\Delta\varphi = \Delta U + \Delta w_{0_{\text{rev}}} - \int T\,dS = -\Delta w_{1_{\text{rev}}} > 0, \tag{8.12}$$

where $\Delta\varphi$ is merely a symbol for the middle terms of Equation (8.12) and does not represent the increment, for example, of some function of state φ. By this we mean that it is the entire symbol $\Delta\varphi$ which has meaning, and that no meaning is attached to φ alone. $\Delta\varphi$ is just a shorthand notation for $\Delta U + \Delta w_{0_{\text{rev}}} - \int T\,dS$. On the other hand, $\Delta\varphi$ is always positive for stable equilibrium, and it is this property which makes it useful for the development of thermodynamic potentials.

The reader should fix in his mind, once again, the fact that the process involved in Equation (8.12) is a *virtual* variation which begins in an equilibrium state determined by the constraints $\mathcal{X}_0$ and ends in another state determined by both $\mathcal{X}_0$ and $\mathcal{X}_1$. Equation (8.12) asserts that for every such variation in which the constraints $\mathcal{X}_0$ are not relaxed, $\Delta\varphi$, defined as it is, increases. We have limited our considerations to variations which take place reversibly. Such variations are the only ones which must be considered in the orderly development and application of thermodynamic potentials.

Notice in particular that $\Delta\varphi$, defined as $\Delta U + \Delta w_{0_{\text{rev}}} - \int T\,dS$ does not contain $-(\Delta w_1)_{\text{rev}}$ explicitly. Thus, as we have stated previously, the explicit natures of the constraint $\mathcal{X}_1$ or of the end state to which the system is displaced do not enter the problem. For this reason, authors of books on

† We are here taking the term "equilibrium" to be synonymous with "stable equilibrium." This means that when the constraint driving the system away from its initial state is removed, the system returns to that state. The same is not true of "*unstable* equilibrium" which is discussed in Chapter 14.

thermodynamics have tended to ignore both $\mathcal{X}_1$ and the displaced state which it produces. This neglect has been carried to such a degree that even the equilibrium nature of the end state has been denied. Thus, one can actually find misleading statements such as, "A virtual variation is so general that it may carry a system from a state of equilibrium to a state of nonequilibrium." Since the explicit nature of the displaced state does not enter the equations, mathematical manipulations may be performed properly even though philosophical comprehension is lacking. In fact, this is precisely the path which most texts on thermodynamics pursue (especially when the author is anxious to focus attention on the applications rather than upon the philosophy of the subject). However, failure to understand the philosophy means failure to understand thermodynamics. This is unfortunate because with so abstract a discipline a thorough understanding is prerequisite to the exploration of new domains.

4. The Sign of $-\Delta w_{1_{rev}}$

One of the characteristics of a reversible process (see Chapter 1, Section 7) is the fact that every one of its component processes is reversed by a reversal of driving force. Thus, in the process inverse to the one to which Equation (8.10) applies, the work performed in connection with $\mathcal{X}_1$ *by* the system is $-\Delta w_{1_{rev}}$, the same quantity which appears in Equation (8.11) where it represents work performed *on* the system.† In demonstrating the sign of $-\Delta w_{1_{rev}}$, it is more convenient to begin in the displaced state determined by both $\mathcal{X}_0$ and $\mathcal{X}_1$ and to examine the process of return to the initial state, determined by $\mathcal{X}_0$ alone. This is the process in which, if it is conducted reversibly, the work performed in connection with $\mathcal{X}_1$ by the system will be $-\Delta w_{1_{rev}}$.

It is clear that if $\mathcal{X}_1$ is removed, a return to the initial state will occur since by definition that state is the one determined by the remaining constraints $\mathcal{X}_0$. The process of return will of course be irreversible, because only through the application of $\mathcal{X}_1$ can the system be maintained in an equilibrium state

† According to the usage introduced in Chapter 1, Section 6, and retained throughout this book, the symbols Dw and Δw, with positive sign, refer to work performed *by* the system. Some confusion may therefore be generated by our use, now, of $-\Delta w_{1_{rev}}$ (with *negative* sign) to denote work performed *by* the system. However, it should be borne in mind that the symbol $\Delta w_{1_{rev}}$ had its origin in Equation (8.11) and is meant to indicate the work done *by* the system in the process *to which* (8.11) *refers*. Since the positive sign is used there for work done *by* the system, the conventional usage is involved. Our use now of $-\Delta w_{1_{rev}}$ for work done *by* the system does not represent a sudden departure from convention, for the process being considered now is the inverse of the one to which (8.11) applies, and $\Delta w_{1_{rev}}$ still refers to work performed *by* the system in *that* process. For consistency, therefore, $-\Delta w_{1_{rev}}$ must be the work performed *by* the system in the inverse process.

other than the initial one; that is, only through the application of X_1 can the process of return be rendered quasistatic and capable of being plotted in state space.

The total work performed by the system in connection with the constraint X_0 during this irreversible return may be denoted by $-\Delta w_0$, where the absence of the suffix rev which appears, for example, in Equation (8.10) indicates that the quantity refers to the change conducted irreversibly. For the same change conducted reversibly, through the intervention of X_1, the work is $-(\Delta w_{0_{\text{rev}}} + \Delta w_{1_{\text{rev}}})$. According to Equation (4.45), the work performed reversibly for a given change exceeds that performed irreversibly so that

$$-(\Delta w_{0_{\text{rev}}} + \Delta w_{1_{\text{rev}}}) + \Delta w_0 > 0. \tag{8.13}$$

If the steps leading to Equation (4.45) are retraced, it will be seen that it (and therefore Equation (8.13) also) is a consequence of Equation (4.2). Thus, it is through Equation (8.13) that the inequality involving the entropy enters the argument. If it were so that

$$\Delta w_{0_{\text{rev}}} = \Delta w_0, \tag{8.14}$$

the proof would be complete, because substitution of Equation (8.14) into Equation (8.13) would yield

$$-\Delta w_{1_{\text{rev}}} > 0, \tag{8.15}$$

which is the same as Equation (8.11). As we shall see below, Equation (8.14) is not true in general, but we observe that $\Delta w_{1_{\text{rev}}}$ is really independent of the precise value of Δw_0 (since Δw_0 refers to the change taking place over one path, and $\Delta w_{1_{\text{rev}}}$ refers to the same thermodynamic change taking place over another path). If, therefore, we can find *even one* irreversible path over which Equation (8.14) holds, we can use that path and the corresponding equation (8.14) together with (8.13) (which is true in general, independent of the precise value of Δw_0) to prove Equation (8.15).

To see how Equation (8.14) is not true in general, but how one may always find an irreversible path over which it *is* true, consider the electrochemical cell of Chapter 1, Section 8. Here we may let Δw_0 refer to the *volume* work performed by the cell while Δw_1 corresponds to *electrical* work. Suppose that when the cell is discharged irreversibly (in the absence of an impressed emf, which when present may be represented by X_1) so that $\Delta w_1 = 0$, the process occurs so rapidly that the accompanying change of volume is almost explosive. Then, nonhydrostatic stresses will be set up in the surrounding atmosphere and Δw_0 will *not* be the simple volume work $p\,\Delta V$ (where p is the atmospheric pressure). On the other hand, when an external emf is impressed across the cell terminals so as to retard the flow of current, the process may be conducted reversibly. Now the volume work, $\Delta w_{0_{\text{rev}}}$, *will be* $p\,\Delta V$; and in addition, electrical work, $\Delta w_{1_{rev}}$, will be performed. Thus, in this example, $\Delta w_{0_{rev}}$ is different from Δw_0 and Equation (8.14) is not true.

Suppose, however, that the conventional atmosphere is replaced by an ambient with infinitely fast relaxation mechanisms, that is, a reversible environment of the sort discussed in Sections 7 and 8 of Chapter 4. Then in spite of the speed of the electrochemical reaction, the surroundings will always behave hydrostatically so that in this case Δw_0 is $p\,\Delta V$. Thus, for a path involving a reversible environment Equation (8.14) can be made valid. Since the thermodynamic states of the system are independent of the nature of the relaxation mechanisms in the surroundings, no loss of generality is incurred by the device of using an artificial fast relaxing environment. It is to be emphasized that an environment which always behaves reversibly *does not* imply that the system itself may only undergo reversible changes. In the example just quoted, it is clear that even in the presence of the fast relaxing surroundings no electrical work is performed, and the cell discharges irreversibly.

The device of the reversible environment may always be employed to insure the validity of Equation (8.14), no matter what kind of work and constraints Δw_0 and χ_0 refer to. Thus, one may always find *at least one* irreversible path, for any given change of state, corresponding to which Equation (8.14) is valid. As explained above, this demands the general validity of Equation (8.15). The proof is therefore complete, and Equation (8.11) (or (8.15)) emerges as a consequence of Equation (4.2).

5. Thermodynamic Potentials and Extremal Conditions

The validity of Equation (8.11) requires Equation (8.12) to be true. This inequality has the following meaning. Suppose that a closed system is in a state of stable equilibrium determined by a set of constraints χ_0. If now a new constraint χ_1 is introduced (while all the original constraints χ_0 are retained) which displaces the system to a new state of *equilibrium* determined by both χ_0 and χ_1, the value of $\Delta\varphi$ corresponding to this displacement is always positive. $\Delta\varphi$ is evaluated by the middle terms of Equation (8.12) which makes no explicit reference to the constraint χ_1 or to the end state to which the system is displaced. This end state, however, having more constraints than the initial state, will also possess more independent variables. The displacement is a "virtual" one because neither χ_1 nor the properties of the end state enter explicitly into the specification of $\Delta\varphi$.

Since $\Delta\varphi$ is positive for any virtual variation in which χ_0 is retained, it is clear that under such conditions $\Delta\varphi$ will have a minimum in the initial state. This suggests that $\Delta\varphi$ may be used to develop a thermodynamic potential.

It is not convenient to use $\Delta\varphi$ itself for this purpose since it is not the increment of some function of state but rather a symbol for the middle terms of Equation (8.12). On the other hand, it may be possible to either find or synthesize functions of state whose increments *imitate* $\Delta\varphi$ over certain *selected* paths in state space. Along one such path, therefore, the particular function of state has a minimum at the initial state, and therefore it may be used as a thermodynamic potential. Since

$$\Delta\varphi = \Delta U + \Delta w_{0_{\mathrm{rev}}} - \int T\,dS, \tag{8.16}$$

it is clear that the functions of state whose increments *imitate* $\Delta\varphi$ will depend upon the form of $\Delta w_{0_{\mathrm{rev}}}$, the reversible work performed in connection with the constraints χ_0; and therefore the thermodynamic potentials which are suitable for a given system will depend upon how that system is constrained or upon the kinds of work which the system is capable of performing (see Chapter 1, Section 8).

Let us begin with a system capable only of volume work. Under this condition,

$$\Delta w_{0_{\mathrm{rev}}} = \int p\,dV, \tag{8.17}$$

so that $\Delta\varphi$ in Equation (8.16) becomes

$$\Delta\varphi = \Delta U + \int p\,dV - \int T\,dS. \tag{8.18}$$

If a path in state space is considered which runs through the initial state, and along which S and V are held constant, we may denote the value of $\Delta\varphi$ on this path as $(\Delta\varphi)_{S,V}$. From Equation (8.18) it may be seen that since $dV = dS = 0$, along this path

$$(\Delta\varphi)_{S,V} = (\Delta U)_{S,V}, \tag{8.19}$$

so that the increment of the internal energy *imitates* $\Delta\varphi$ over a path of constant entropy and volume. Therefore, U may be employed as a thermodynamic potential along such a path.

If instead of a path of constant S and V we consider one for which S and p are constant, Equation (8.18) becomes

$$(\Delta\varphi)_{S,p} = \Delta U + p\,\Delta V. \tag{8.20}$$

Now consider the increment of the enthalpy H defined by Equation (3.20) along this path. We have

$$(\Delta H)_{S,p} = \Delta U + p\,\Delta V. \tag{8.21}$$

Comparison of the right sides of Equations (8.20) and (8.21) shows that

$$(\Delta\varphi)_{S,p} = (\Delta H)_{S,p}, \tag{8.22}$$

so that along a path of constant entropy and pressure the increment of H *imitates* $\Delta\varphi$, and H may be used as a thermodynamic potential.

Next consider a path of constant temperature and volume. By holding T and V constant in Equation (8.18) we get

$$(\Delta\varphi)_{T,V} = \Delta U - T\,\Delta S. \tag{8.23}$$

The increment of the Helmholtz free energy A defined in Equation (4.67) along this path is

$$(\Delta A)_{T,V} = \Delta U - T\,\Delta S, \tag{8.24}$$

and comparison of the right sides of Equations (8.23) and (8.24) reveals that

$$(\Delta\varphi)_{T,V} = (\Delta A)_{T,V}, \tag{8.25}$$

so that the increment of A *imitates* $\Delta\varphi$ and A may be employed as a thermodynamic potential along a path for which T and V are fixed.

Finally, consider a path of constant temperature and pressure. Along this path Equation (8.18) becomes

$$(\Delta\varphi)_{T,p} = \Delta U + p\,\Delta V - T\,\Delta S. \tag{8.26}$$

Along the same path we have for the increment of the Gibbs free energy G, defined by Equation (4.71),

$$(\Delta G)_{T,p} = \Delta U + p\,\Delta V - T\,\Delta S. \tag{8.27}$$

Comparison of these two equations shows that

$$(\Delta\varphi)_{T,p} = (\Delta G)_{T,p}, \tag{8.28}$$

so that along a path of constant T and p the increment of G *imitates* $\Delta\varphi$ and G may be used as a thermodynamic potential.

The extremal conditions (comparable to Equation (7.1)) for the closed system capable of volume work only are then,

$$(\delta U)_{S,V} = 0, \tag{8.29}$$
$$(\delta H)_{S,p} = 0, \tag{8.30}$$
$$(\delta A)_{T,V} = 0, \tag{8.31}$$
$$(\delta G)_{T,p} = 0. \tag{8.32}$$

Actually, because these relations pertain to a closed system, the subscripts should include the symbol n to indicate the constancy of the mole numbers.

However, it has been conventional, in thermodynamics, to regard such constancy as having been understood, and in this book we will not upset this convention.

We might write Equation (8.16) in differential form, corresponding to Equations (8.29) through (8.32). Thus,

$$D\varphi = \delta U + Dw_{0_{\text{rev}}} - T\,\delta S = 0, \tag{8.33}$$

where $D\varphi$ is not the differential of some function of state φ, but is simply shorthand for the middle terms in Equation (8.33). Just as it is only the entire symbol $\Delta\varphi$ which has meaning, so it is only the entire symbol $D\varphi$ which has meaning.

Equation (8.33) is a generalized extremal condition for all *closed* systems with uniform temperature. It symbolizes the fact that $\Delta\varphi$ has an extremal value in the initial state if the constraints χ_0 are retained.

6. Generalized Feature of Equilibrium

Equations (8.29) through (8.32) may be regarded as "features" rather than "criteria" of equilibrium in much the same sense that Equation (7.1) (see Chapter 7, Section 3) is a "feature" rather than a "criterion" of equilibrium. However, these extremal conditions, as in the case of (7.1), contain less information than expressions like Equations (8.19), (8.22), (8.25), and (8.28) which, in conjunction with (8.12), require U, H, A, and G to have *minima* in the initial state. Equations (8.29) through (8.32) would remain valid if there were maxima as well as minima. The question of minima or maxima in the present context is connected with the problem of "stability," and we shall defer discussion of this point until Chapter 14. It may be added, however, that "features" of equilibrium embodied in Equations (8.29) through (8.32) are equivalent to the "elementary method" described in Section 2 of this chapter. Since that method only requires a state to be representable in state space, it applies to both stable and unstable equilibria, and therefore the same is true of the "equivalent" methods contained in Equations (8.29) through (8.32). We shall have more to say about this later.

It is absolutely necessary to emphasize that all the thermodynamic potentials U, H, A, and G, as well as the features of equilibrium embodied in their differentials taken along selected paths, are logically equivalent. They are all features of equilibrium *for one and the same initial state*. It is true that the end states of the displacements (taken over different paths) are in each case different, but this is of no consequence since the details of these end states are not in question. We are interested in the nature of the system in its *initial state* and this is the same in each case.

The same equivalence extends to S when it is used as a thermodynamic potential. The initial state may be the same one to which U, H, A, and G are applied along with their respective paths. The special path for S is the inverse of one over which the system could have reached the initial state while isolated. It is especially important to emphasize this equivalence of S because most texts view S as a thermodynamic potential suitable for an *isolated* system while U, H, A, and G are often regarded as thermodynamic potentials for *nonisolated* systems. The fallacy in this point of view is of course exposed by the above-mentioned equivalence, and it should be clear that the concepts of isolation enter the argument only through the special characteristics of the path over which S may be employed as a thermodynamic potential, and then only through the properties of inverseness. The actual path is *not* one which may be traversed if the system remains isolated.

All of the features of equilibrium described in this chapter can be used for the study of internal equilibrium in systems *known to be at equilibrium.* They will be used for this purpose in what follows. It is important to remember that they have been derived for closed systems limited to the performance of volume work in connection with the constraints χ_0, and that none of them is to be preferred logically to the other. Often too much emphasis is placed on the Gibbs free energy as a thermodynamic potential. Consequently, many students have come to feel that there is some special fundamental connection between G and equilibrium.

Actually, the only fundamental relation is the generalized "feature" of equilibrium (8.33). This is applicable to any closed system, constrained in any manner, and able to perform any kind of work. Confusion in the application of thermodynamics will be avoided if the treatment of each new system is referred to Equation (8.33). In each new case, it will be necessary to find (sometimes by synthesis) functions of state whose differentials will *imitate* $D\varphi$ over selected paths in state space. The form of $D\varphi$ will vary, depending upon the constraints χ_0 and therefore upon the kind of work which the system may perform.

It is useful to express $\Delta\varphi$ in terms of functions of state so that the coefficients of complete differentials may be examined and useful relationships such as Equation (7.24) may be derived. However, there are other advantages in this procedure. Thus, consider a system to which a constraint has been applied to drive it away from an initial equilibrium state over a path of constant T and p. The new equilibrium state to which the system has been driven will obviously possess a larger value of G since G is minimized in the initial state along a path of constant T and p. Since G is a function of state, it is possible, in principle, to measure and to tabulate its values in the displaced state and the original state (both of course at the same temperature and pressure).

Suppose the system is initially in the displaced state. Then the removal of the constraint will permit it to undergo a spontaneous transition (or "induced" transition, see Chapter 4, Section 2) to what we have been calling the initial state. A comparison of the tabulated values of G in both states will then reveal that $\Delta G < 0$ for this change. Thus, if a system possesses two states at the same temperature and pressure, it may undergo a spontaneous transformation from the state of higher G to that of lower G. It is clear that a constraint must be removed in the state of higher G before the transformation may occur, and that the state of lower G will possess fewer independent variables.

Since many thermodynamic processes take place in laboratory thermostats, they represent changes between states with the same values of T and p. Thus, tabulated values of G can be employed to predict whether such changes will occur upon the removal of a constraint. For this reason more attention has been devoted to G than to other thermodynamic potentials.

However, the prediction of the occurrence or nonoccurrence of processes is only a small part of thermodynamics, whose more important service remains the derivation of interconnections between macroscopic variables. For this purpose, all the thermodynamic potentials are useful and many of them are logically equivalent.

7. Alternative Representations of the Chemical Potential

The Legendre transformations performed in Chapter, 4 Section 13 were all based upon the use of Equation (4.74) which is only valid for a *closed* system. For an *open* system, such as one of the phases of the two-phase system of Chapter 7, Section 2, Equation (4.74) must be replaced by Equation (7.3) which may be written (without superscripts) in the form

$$dU = T\,dS - p\,dV + \mu\,dn, \tag{8.34}$$

where μ, the *chemical potential*, is (as in the case of Equation (7.4)

$$\mu = \left[\frac{\partial U}{\partial n}\right]_{S,V}. \tag{8.35}$$

Now if Equation (8.34) is used in place of (4.74) in Chapter 4, Section 13, we derive the following expressions:

$$dH = T\,dS + V\,dp + \mu\,dn, \tag{8.36}$$

$$dA = p\,dV - S\,dT + \mu\,dn, \tag{8.37}$$

$$dG = V\,dp - S\,dT + \mu\,dn. \tag{8.38}$$

By identifying the coefficients in Equations (8.34), (8.36), (8.37), and (8.38), in the manner of Equation (2.4), we obtain

$$T = \left[\frac{\partial U}{\partial S}\right]_{V,n} = \left[\frac{\partial H}{\partial S}\right]_{p,n}, \tag{8.39}$$

$$p = -\left[\frac{\partial U}{\partial V}\right]_{S,n} = -\left[\frac{\partial A}{\partial V}\right]_{T,n}, \tag{8.40}$$

$$V = \left[\frac{\partial H}{\partial p}\right]_{S,n} = \left[\frac{\partial G}{\partial p}\right]_{T,n}, \tag{8.41}$$

$$S = -\left[\frac{\partial A}{\partial T}\right]_{V,n} = -\left[\frac{\partial G}{\partial T}\right]_{p,n}, \tag{8.42}$$

$$\mu = \left[\frac{\partial U}{\partial n}\right]_{S,V} = \left[\frac{\partial H}{\partial n}\right]_{S,p} = \left[\frac{\partial A}{\partial n}\right]_{T,V} = \left[\frac{\partial G}{\partial n}\right]_{T,p}. \tag{8.43}$$

Equation (8.43) illustrates alternative methods for measuring μ in one and the same state. According to the last term in (8.43), μ may be measured (in principle) by starting with the system, in the state in which μ is to be evaluated, and adding dn moles in such a manner that T and p remain constant. The ratio of dG, the corresponding change in G, to (in the limit as $dn \to 0$) dn then measures μ. On the other hand, according to the second term in Equation (8.43), we may start with the system in the state in which μ is to be measured and add dn moles in such a manner that S and V remain constant. The ratio of dU to dn for this change also measures μ.

In both cases the value of the chemical potential μ is measured in the same state. However, the state to which the system is displaced, for the measurement, is different in each case, since in one instance a constant T, p path in state space is involved and in the other a constant S, V path. In both cases the displacement results in the addition of dn moles, but the constraints χ_1, which are applied to force the variations, must be different to assure that the system pursues the proper respective paths. As in our previous considerations, neither the precise natures of the constraints nor the end states of the variations need be specified. In the limit as $dn \to 0$ it is the value of μ in the *initial* state which is measured. Both displacements begin at the same point in state space (same initial state) but the variations follow different paths.

In practice, μ is not measured by performing either of these variations, but more convenient techniques are employed. However, if one were forced to perform the measurement using the last or the second terms of Equation (8.43), it is clear that the last term would be more convenient. This follows from the fact that it is easier to produce a path of constant temperature and pressure in the laboratory than one of constant entropy and volume. The last definition of μ in (8.43) seems to have been given more prominence than the definition contained in the second term, and again G seems to play a more fundamental role than U or the other thermodynamic potentials. Part of the reason for this unfounded point of view lies in the erroneous identification of logic with convenience. It must be emphasized again that both definitions are logically equivalent. In fact, other definitions are possible as is evidenced by the third and fourth terms of Equation (8.43).

8. Proof of the Conditions of Internal Equilibrium using either U, H, A, or G

Equation (8.43) provides alternative definitions of the chemical potential μ. It is of interest to prove Equation (7.24) using U, H, A, or G. We have already proved this relation for a single-component, two-phase system by the elementary method of Section 2 of this chapter.

Either of the thermodynamic potentials H, A, or G may be used for the proof, so let us begin with G. As in the case of S in Chapter 7, G may be decomposed into two terms, one for each phase.

$$G = G^{(1)} + G^{(2)}. \tag{8.44}$$

We assume that thermal and mechanical equilibria have been established so that temperature and pressure are uniform throughout the system, or else Equation (8.32) will be inapplicable since it has been derived for systems throughout which both T and p have unique values.† Substitution of Equation (8.44) into Equation (8.32) then yields

$$(\delta G)_{T,p} = (\delta G^{(1)})_{T,p} + (\delta G^{(2)})_{T,p} = 0. \tag{8.45}$$

With temperature and pressure constant, Equation (8.38) becomes, for each phase,

$$\begin{aligned} dG^{(1)} &= \mu^{(1)}\, dn^{(1)}, \\ dG^{(2)} &= \mu^{(2)}\, dn^{(2)}, \end{aligned} \tag{8.46}$$

where superscripts have now been used to distinguish the phases. By substituting these relations into Equation (8.45) we get

$$\mu^{(1)}\, \delta n^{(1)} + \mu^{(2)}\, \delta n^{(2)} = 0, \tag{8.47}$$

† For example, if each phase had a different pressure, the work performed by the system would not be represented by $p\, dV$ but by $p^{(1)}\, dV^{(1)}$ and $p^{(2)}\, dV^{(2)}$, and Equation (8.17) would no longer be valid. This merely re-emphasizes the fact that the thermodynamic potentials are *characteristic* of the kind of work which the system may perform in connection with the constraints X_0.

and since the system is closed,

$$\delta n^{(2)} = -\delta n^{(1)}. \tag{8.48}$$

Introduction of this into Equation (8.47) yields

$$(\mu^{(1)} - \mu^{(2)})\,\delta n^{(1)} = 0 \tag{8.49}$$

or

$$\mu^{(1)} = \mu^{(2)}.$$

One may demonstrate the same result by employing Equations (8.30) and (8.31). Thus, with (8.30) we have

$$(\delta H)_{S,p} = (\delta H^{(1)} + \delta H^{(2)})_{S,p} = 0. \tag{8.50}$$

Now it is also true that

$$S = S^{(1)} + S^{(2)} \qquad \text{and} \qquad dS = dS^{(1)} + dS^{(2)}. \tag{8.51}$$

By introducing Equation (8.36) in the forms

$$\begin{aligned} dH^{(1)} &= T\,dS^{(1)} + V^{(1)}\,dp + \mu^{(1)}\,dn^{(1)}, \\ dH^{(2)} &= T\,dS^{(2)} + V^{(2)}\,dp + \mu^{(2)}\,dn^{(2)}, \end{aligned} \tag{8.52}$$

we obtain

$$(\delta H)_{S,p} = \mu^{(1)}\,\delta n^{(1)} + \mu^{(2)}\,\delta n^{(2)} = 0, \tag{8.53}$$

where both Equation (8.51) and the condition of constant p has been applied. The equation on the right of (8.53) is identical with Equation (8.47), and because Equation (8.48) holds the result, Equation (8.49), is obtained once again.

The use of Equation (8.31) yields

$$(\delta A)_{T,V} = (\delta A^{(1)} + \delta A^{(2)})_{T,V} = 0. \tag{8.54}$$

Here we remember that

$$V = V^{(1)} + V^{(2)} \qquad \text{and} \qquad dV = dV^{(1)} + dV^{(2)}. \tag{8.55}$$

Substitution of Equation (8.37) into Equation (8.54), using both Equation (8.55) and the condition of constant T, once again yields Equation (8.47); and this, together with Equation (8.48), produces Equation (8.49) which is the same as Equation (7.24).

Thus, as might have been expected, all the thermodynamic potentials yield the same result. The respective variations have been carried out over *different* selected paths in state space, but they all initiate at the *same* point and provide information about the condition of internal equilibrium in the *one* state to which this initial point corresponds.

Notice that Equation (8.49), which is really a direct rendering of the extremal condition (8.32) (or for that matter, of Equations (8.29), (8.30), and (8.31)), is identical with Equation (8.8) which is the fundamental equation of the "elementary" method described in Section 2. Thus, the features of equilibrium embodied in Equations (8.29) through (8.32) are indeed equivalent to the equations of the elementary method. They are therefore valid for all kinds of equilibrium (the only restriction being that the state of equilibrium be representable as a point in state space) and make no distinction between *stable* and *unstable* equilibria.

9. Generalization to any Number of Phases with any Number of Components

The results of the preceding sections may be generalized to a multiphase, multicomponent system. In dealing with more complicated situations, a suitable notation must be employed. The following conventions will be adopted. Superscripts will be employed as usual to denote phases and subscripts to denote components. Thus, $n_i^{(j)}$ symbolizes the number of moles of the ith species in the jth phase. Similarly, $V^{(j)}$ is the volume of the jth phase, et cetera.

For the jth phase with C components, Equation (8.38) generalizes to the following form:

$$dG^{(j)} = V^{(j)}\,dp - S^{(j)}\,dT + \sum_{i=1}^{i=C} \mu_i^{(j)}\,dn_i^{(j)}, \tag{8.56}$$

where

$$\mu_i^{(j)} = \left[\frac{\partial G^{(j)}}{\partial n_i^{(j)}}\right]_{T,p} \tag{8.57}$$

is the chemical potential of the ith component in the jth phase. In Equation (8.57) we have followed the convention adopted in the paragraph following Equation (5.66) of Chapter 5, Section 7. Thus, the subscripts in Equation (8.57) should really contain the mole numbers of all species in the phase except species i. The notation is very much simplified if these mole numbers are omitted. No confusion should result, since it will always be clear when such an omission has been made.

As in the case of a one-component system, $\mu_i^{(j)}$ may be defined in terms of U, H, and A as well as G (see Equation (8.43)).

If there are P phases, then

$$G = \sum_{j=1}^{j=P} G^{(j)} \tag{8.58}$$

and

$$dG = \sum_{j=1}^{j=P} dG^{(j)}. \tag{8.59}$$

If Equation (8.32) is applied so that the uniform temperature and pressure remain constant during the variation, the first two terms on the right of Equation (8.56) vanish; and substitution of the result into Equation (8.59) and of (8.59) into (8.32) yields

$$(\delta G)_{T,p} = \sum_j \sum_i \mu_i^{(j)} \, \delta n_i^{(j)} = 0. \tag{8.60}$$

In this equation we have used the notation $\sum\limits_j \sum\limits_i$ in place of the more ponderous symbolism $\sum\limits_{j=1}^{j=P} \sum\limits_{i=1}^{i=C}$. No confusion should result from simplification, especially because superscripts will always denote phases while subscripts will be used for components. We will follow this practice from now on. Since the system is closed, the following relations are also valid:

$$\sum_j \delta n_i^{(j)} = 0, \qquad i = 1, 2, \ldots, C. \tag{8.61}$$

Equation (8.60), together with the set of relations (8.55), constitutes an example of the problem solved in Chapter 2, Section 6. In fact, the relations (8.61) are of the same form as the set (2.48), if the derivatives $\dfrac{\partial \alpha_1}{\partial x}$, $\dfrac{\partial \alpha_2}{\partial y}$, et cetera, are placed equal to zero or unity in such a way as to make the parallelism complete. Equation (8.60) is the counterpart of (2.31), with $\mu_i^{(j)}$ taking the place of the derivatives $\dfrac{\partial F}{\partial x}$, and so forth.

We proceed in the same manner as in Chapter 2, Section 6. Each of the relations (8.61) is multiplied by an undetermined parameter λ_i, and all are added to Equation (8.60). The result is

$$\sum_i \sum_j (\mu_i^{(j)} + \lambda_i) \, \delta n_i^{(j)} = 0. \tag{8.62}$$

Since, as in Chapter 2, each coefficient must be zero, we arrive at the result

$$\mu_i^{(j)} = \lambda_i. \tag{8.63}$$

Thus, at equilibrium the chemical potential of the ith species is the same in every phase, that is, λ_i does not depend upon j. This is the general result we have been seeking.

10. The Phase Rule

The *phase rule* in its conventional specialized form provides a relation between the so-called intensive *degrees of freedom* or *variance* f of a system and the number of phases P and the number of independently variable components C. The usual relation is

$$f = C - P + 2. \tag{8.64}$$

This relation is discussed in most texts on physical chemistry, and we shall not elaborate the subject here. The form (8.64) is limited to systems in which the only variables of state besides variables of composition are temperature and pressure, both presumed uniform throughout the system.

The variance f is the number of *intensive* variables in the system which can be varied without changing the number of phases. Hence, the term "phase rule."

In this section we merely wish to demonstrate that Equation (8.64) follows from the results of Section 9. To accomplish this we note that if a given phase has C components there will be $C - 1$ independent *concentrations* or *intensive* variables of composition in that phase.† If there are P phases there will be $P(C - 1)$ intensive variables of composition in the system. Together with temperature and pressure then, there will be

$$P(C - 1) + 2 \tag{8.65}$$

intensive variables. However, Equation (8.63), which may be written in the form

$$\mu_i^{(1)} = \mu_i^{(2)} = \mu_i^{(3)} = \cdots \mu_i^{(P)} \qquad i = 1, 2, \ldots, C, \tag{8.66}$$

constitutes a set of $C(P - 1)$ relations among the intensive variables enumerated in Equation (8.65), since the $\mu_i^{(j)}$ are functions of these variables. Thus, f, the number of independent intensive variables, is Expression (8.65) reduced by $C(P - 1)$.

$$f = P(C - 1) + 2 - C(P - 1) = C - P + 2, \tag{8.67}$$

which is just the phase rule (8.64).

The 2 in Equation (8.64) or Equation (8.67) appears because only temperature and pressure have been considered (besides variables of composition) as intensive variables, and furthermore, because the pressure is considered the same in all phases. If additional intensive variables existed, or if the pressure were not uniform throughout the system, some other number besides 2 would appear. The reader is referred to any text on physical chemistry and to most of the texts on thermodynamics recommended in the preface for a more detailed discussion of the phase rule. The meaning of the quantity C, the number of *independently* variable components, is especially important and is discussed in the references cited.

11. Chemical Potential as a Partial Molar Quantity

When the definition (8.57) of the chemical potential μ_i (dropping the phase index j, since we will now focus on one phase) is compared with the

† For example, the mole fraction x_i is an intensive variable of composition and there always exists the relation $\sum_i x_i = 1$ among the x_i's. Thus, if there are C x_i's only $C - 1$ of them are independent.

definition of partial molar quantities given in Chapter 3, Section 9, it becomes clear that μ_i is the partial molar Gibbs free energy of the ith species. Thus,

$$\mu_i = \bar{G}_i. \tag{8.68}$$

Thus, in accordance with Equation (3.30) (or by application of Euler's theorem (2.43)), we have

$$G = \sum_i n_i \mu_i. \tag{8.69}$$

A relationship analogous to Equation (3.34) may be derived immediately. Thus, the total differential of Equation (8.69) is

$$dG = \sum_i n_i \, d\mu_i + \sum_i \mu_i \, dn_i. \tag{8.70}$$

Now subtracting Equation (8.56) from Equation (8.70) (dropping the j) yields

$$S\,dT - V\,dp + \sum_i n_i \, d\mu_i = 0. \tag{8.71}$$

This rather famous equation has been called the Gibbs-Duhem relation.

Notice that despite the fact that μ can be defined in terms of the partial derivatives of any of the thermodynamic potentials with respect to the mole number (as in Equation (8.43)), it is only in the case of the Gibbs free energy that the derivative is taken with T and p constant. Hence, chemical potential cannot be expressed in terms of partial molar quantities of the other thermodynamic potentials,

$$\begin{aligned} \mu_i &\neq \bar{U}_i \\ &\neq \bar{H}_i \\ &\neq \bar{A}_i. \end{aligned} \tag{8.72}$$

Equation (5.64) for the partial molar Gibbs free energy of a component of an ideal solution may now be written in the form

$$\overset{0}{\mu_i} = \overset{0}{\mu_i}(T,p) + RT \ln x_i, \tag{8.73}$$

where $\mu_i^0(T,p)$ is the chemical potential of pure component i.

12. Open Systems

Thermodynamic potentials may even be derived for *open* systems. For such systems Equation (8.10) must be replaced by

$$\Delta U = \int T\,dS - \Delta w_{0\text{rev}} - \Delta w_{1\text{rev}} + \sum_i \sum_j \int \mu_i^{(j)} \, dn_i^{(j)}, \tag{8.74}$$

where the sums over i and j extend over C components and P phases, and where the last term accounts for the change in internal energy which may now occur because of an exchange of matter with the surroundings. Equation (8.74) may be rearranged (as in the case of Equation (8.12)) to yield

$$\Delta\varphi_{\text{open}} = \Delta U + \Delta w_{0_{\text{rev}}} - \int T\,dS - \sum_i \sum_j \int \mu_i^{(j)}\,dn_i^{(j)} = -\Delta w_{1_{\text{rev}}} > 0. \tag{8.75}$$

In this expression, $\Delta\varphi_{\text{open}}$ is merely a symbol for the middle terms as is $\Delta\varphi$ in Equation (8.12). Furthermore, the inequality on the right may be proved using the same technique employed for closed systems in Section 4 of this chapter.

Just as in the case of closed systems, we seek functions of state whose increments *imitate* $\Delta\varphi_{\text{open}}$. As an example, assume that the system is capable only of volume work so that $\Delta w_{0_{\text{rev}}} = \int p\,dV$. Then,

$$\Delta\varphi_{\text{open}} = \Delta U + \int p\,dV - \int T\,dS - \sum_i \sum_j \int \mu_i^{(j)}\,dn_i^{(j)}. \tag{8.76}$$

If we choose a path in state space such that T, V, and all the $\mu_i^{(j)}$ are held constant, and denote the constancies of all the μ's by the single subscript μ,

$$(\Delta\varphi_{\text{open}})_{T,V,\mu} = \Delta U - T\Delta S - \sum_i \sum_j \mu_i^{(j)}\,\Delta n_i^{(j)}. \tag{8.77}$$

Now consider the function of state $-pV$. According to Equations (4.71) and (8.69),

$$-pV = -G + U - TS = -\sum_i \sum_j \mu_i^{(j)} n_i^{(j)} + U - TS. \tag{8.78}$$

Furthermore,

$$[\Delta(-pV)]_{T,V,\mu} = -\sum_i \sum_j \mu_i^{(j)}\,\Delta n_i^{(j)} + \Delta U - T\Delta S. \tag{8.79}$$

Comparison of the right sides of Equation (8.79) and Equation (8.77) shows that

$$[\Delta(-pV)]_{T,V,\mu} = (\Delta\varphi_{\text{open}})_{T,V,\mu} > 0, \tag{8.80}$$

so that $-pV$ may be used as a thermodynamic potential for the open system along a path of constant T, V, and μ.

In this book, however, we shall not use thermodynamic potentials for open systems since everything we wish to accomplish may be achieved by restricting attention to closed systems.

IX

Phase Equilibria in Simple Systems

1. Dependence of Chemical Potential on Variables of State

Equation (8.63), which establishes the uniformness of chemical potential throughout an equilibrium system, is of little value unless the relationship of μ to other variables such as temperature, pressure, and composition are known. Unfortunately, this relationship cannot be arrived at by thermodynamic reasoning alone, and at this point further progress demands the introduction of extra-thermodynamic information. The dependence of μ upon other variables may be obtained from theoretical considerations based upon molecular theory or it may be determined by experiment, and both of these methods exceed the limits of normal thermodynamic reasoning.

As we have mentioned earlier, a reasonable approximation for many solutions is to assume the property of ideality. In this case, we may utilize Equation (8.73) to express the chemical potential of the ith component in the jth phase as

$$\mu_i^{(j)} = \mu_i^{(j)0}\,(T,p) + RT \ln x_i^{(j)}. \tag{9.1}$$

For a component of an ideal *gas* solution even a simpler mode of expression exists. Thus, apply Equation (9.1) to a gas. Replace $\bar{G}_i$ in Equation (5.66) by $\mu_i^{(j)}$ and then substitute Equation (9.1). The result is

$$\left[\frac{\partial \mu_i^{(j)}}{\partial p}\right]_{T,n} = \left[\frac{\partial \mu_i^{(j)0}}{\partial p}\right]_{T,n} = V_i^{(j)0} = \frac{RT}{p}. \tag{9.2}$$

Integration of Equation (9.2) yields

$$\mu_i^{(j)0}(T,p) = K_i^{(j)}(T) + RT \ln p, \tag{9.3}$$

where $K_i^{(j)}(T)$ is a function of T alone.

From Equations (5.41), (5.42), and (5.43) it is apparent that for a binary mixture of ideal gases,

$$\begin{aligned} x_1^{(j)} &= \frac{n_1^{(j)}}{n_1^{(j)} + n_2^{(j)}} = \frac{p_1}{p_1 + p_2} = \frac{p_1}{p}, \\ x_2^{(j)} &= \frac{n_2^{(j)}}{n_1^{(j)} + n_2^{(j)}} = \frac{p_2}{p_1 + p_2} = \frac{p_2}{p}. \end{aligned} \tag{9.4}$$

It is evident that, in general, for any multicomponent mixture of ideal gases,

$$x_i^{(j)} = \frac{p_i}{p}. \tag{9.5}$$

Substitution of Equation (9.5) into Equation (9.1) yields

$$\mu_i^{(j)} = \mu_i^{(j)0}(T,p) - RT\ln p + RT\ln p_i. \tag{9.6}$$

If Equation (9.3) is substituted into Equation (9.6) the result is

$$\mu_i^{(j)} = K_i^{(j)}(T) + RT\ln p_i, \tag{9.7}$$

which demonstrates that the chemical potential of a component of an ideal gas solution may be expressed as the sum of a function of temperature alone and RT times the logarithm of the partial pressure. This is to be contrasted with Equation (9.1) where the expression is in terms of the mole fraction $x_i^{(j)}$. There the additive term $\mu_i^{(j)0}$ depends upon both temperature and pressure.

Whenever the values of the variables of state, temperature, pressure, and composition are changed, the system will persist in equilibrium only if the equal chemical potentials continue to be equal. Thus, for example, in a two-phase system where we have $\mu_i^{(1)} = \mu_i^{(2)}$ for all i, the maintenance of this relation requires

$$d\mu_i^{(1)} = d\mu_i^{(2)}, \qquad i = 1, 2, \ldots, C, \tag{9.8}$$

as the magnitudes of the state variables change.

In view of Equation (9.8), there is reason to express $d\mu_i^{(j)}$ in terms of the differentials of the independent variables. Thus,

$$\begin{aligned} d\mu_i^{(j)} &= \left[\frac{\partial \mu_i^{(j)}}{\partial p^{(j)}}\right]_{T,n} dp^{(j)} + \left[\frac{\partial \mu_i^{(j)}}{\partial T}\right]_{p^{(j)},n} dT + \sum_k \left[\frac{\partial \mu_i^{(j)}}{\partial n_k^{(j)}}\right]_{T,p} dn_k^{(j)} \\ &= \bar{V}_i^{(j)}\, dp^{(j)} - \bar{S}_i^{(j)}\, dT + \sum_k \left[\frac{\partial \mu_i^{(j)}}{\partial n_k^{(j)}}\right]_{T,p} dn_k^{(j)}, \end{aligned} \tag{9.9}$$

where we have used Equations (5.66) and (5.67).

In this chapter, Equation (9.8) will be used to derive several useful relations pertaining to phase equilibria in systems capable of volume work only. Our treatment will not be exhaustive since we only wish to illustrate the technique. For more complete coverage the reader is referred to the texts cited in the Preface.

2. The Clapeyron-Clausius Equation

A familiar and important relation concerns the manner in which the vapor pressure of a liquid depends upon temperature. This expression, which is called the Clapeyron-Clausius equation, will now be derived, beginning with Equation (9.8). We consider a one-component, two-phase system consisting of a liquid and its vapor. In place of (j) in Equation (9.1) or of (1) or (2) in Equation (9.8), we now employ (v) and (l) as superscripts to denote the vapor and liquid phases, respectively.

By specializing Equation (9.9) for a one-component phase we get

$$d\mu^{(j)} = d\mu^{(j)0} = V^{(j)0}\,dp^{(j)} - S^{(j)0}\,dT, \tag{9.10}$$

where $V^{(j)0}$ and $S^{(j)0}$ are the molar volumes and entropy, respectively, where the composition terms have been set equal to zero—as they must be in the case of a single component—when it is considered that $\mu^{(j)0}$ refers to the free energy *per mole* and therefore cannot depend upon the number of moles present.

By employing Equation (9.8) and making use of Equation (9.10), together with the superscripts (l) and (v), we obtain

$$V^{(v)0}\,dp - S^{(v)0}\,dT = V^{(l)0}\,dp - S^{(l)0}\,dT, \tag{9.11}$$

where we assume that the pressure is the same in both phases. This equation may be rearranged to yield

$$(V^{(v)0} - V^{(l)0})\,dp = (S^{(v)0} - S^{(l)})\,dT. \tag{9.12}$$

Now

$$S^{(v)0} - S^{(l)0} \tag{9.13}$$

represents the change of entropy experienced by the two-phase system when one mole of substance is transferred from the liquid to the vapor phase at the temperature T and pressure p. This is also the heat absorbed *reversibly* divided by T during this process. Thus,

$$S^{(v)0} - S^{(l)0} = \frac{(\Delta q)_{\text{rev}}}{T}. \tag{9.14}$$

Since the process takes place at constant pressure, we may utilize the discussion of Chapter 3, Section 6 to identify the heat absorbed with ΔH. Thus,

$$S^{(v)0} - S^{(l)0} = \frac{(\Delta H)}{T} = \frac{\lambda_v}{T}, \tag{9.15}$$

where λ_v is the heat of vaporization per mole, equal to the change in enthalpy for every mole of substance vaporized.

Substitution of Equation (9.15) into Equation (9.12) yields,

$$\frac{dp}{dT} = \frac{\lambda_v}{T(V^{(v)0} - V^{(l)0})}. \tag{9.16}$$

This is the exact form of the famous Clapeyron-Clausius equation, which describes how the vapor pressure of a liquid depends upon temperature.

If the vapor is treated as an ideal gas, Equation (5.1) may be applied to $v^{(v)0}$ with the result

$$v^{(v)0} = \frac{RT}{p}. \tag{9.17}$$

Furthermore, if it is assumed (as is usually the case) that

$$v^{(v)0} \gg v^{(l)0}, \tag{9.18}$$

the volume difference in the denominator of Equation (9.16) may be approximated by $v^{(v)0}$ alone. This substitution leads to the familiar relation

$$\frac{d \ln p}{dT} = \frac{\lambda_v}{RT^2}. \tag{9.19}$$

The Clapeyron-Clausius Equation (9.16) may be applied equally well to any phase transition in a one-component system if only λ_v is replaced by the appropriate isobaric heat of transition λ and $v^{(v)0} - v^{(l)0}$ is replaced by Δv, the corresponding change in molar volume. For a condensed phase transition, however, the equation is of most utility in the form

$$\frac{dT}{dp} = \frac{T\Delta v}{\lambda},$$

which gives the change in transition temperature with pressure.

3. Raoult's Law

According to *Raoult's Law*, the ratio of the partial vapor pressure p_i of component i of a solution to the vapor pressure p_i^0 of the pure component at the same temperature equals the mole fraction $x_i^{(l)}$ of component i. It may be proved that this is a direct consequence of the ideal gas and solution laws, and is therefore essentially exact for such systems under the condition that the *total* pressure over the solution is maintained equal to p_i^0 even though p_i itself is not.

For a liquid and gas in equilibrium we may write from Equation (8.63)

$$\mu_i^{(l)} = \mu_i^{(v)}, \tag{9.20}$$

where $\mu_i^{(l)}$ is the chemical potential of the ith species in the liquid and $\mu_i^{(v)}$ is the same quantity in the vapor. If we assume that both the gas solution and the liquid solution are ideal, we can substitute Equation (9.1) for $\mu_i^{(l)}$ and Equation (9.3) for $\mu_i^{(v)}$. The result is

$$\mu_i^{(l)0}(T,p) + RT\ln x_i^{(l)} = K_i^{(v)}(T) + RT\ln p_i$$

or

$$RT\ln \frac{p_i}{x_i^{(l)}} = \mu_i^{(l)0}(T,p) - K_i^{(v)}(T). \tag{9.21}$$

The clearing of logarithms yields

$$\frac{p_i}{x_i^{(l)}} = \exp\left\{\frac{[\mu_i^{(l)0}(T,p) - K_i^{(v)}(T)]}{RT}\right\} = h_i(T,p), \tag{9.22}$$

where h_i is independent of $x_i^{(l)}$. When $x_i^{(l)} = 1$, $p_i = p_i^0$, the vapor pressure of pure i. If the total pressure above the ideal solution is $p = p_i^0$, it is clear that $h_i(T,p_i^0)$ must then be equal to p_i^0. On the other hand, in most experimental situations it happens that $\mu_i^{(l)0}(T,p)$ is not very sensitive to p so that $h_i(T,p) \approx h_i(T)$. Under this circumstance, the following is a good approximation, independent of the exact value of the total pressure:

$$\frac{p_i}{x_i^{(l)}} = p_i^0(T). \tag{9.23}$$

This is Raoult's Law.

4. Boiling Point Elevation and Freezing Point Depression

Boiling is a phenomenon which takes place when the vapor pressure of a liquid equals the pressure exerted by its environment. When the environmental pressure remains fixed, a change in solution composition which produces a change in vapor pressure must lead to a change in boiling point. This is the change in temperature T required to counter-balance the composition change and maintain the equilibrium pressure at the environmental value.

The treatment of this phenomenon is much simplified if one of the components, say the solute, is involatile, that is, exists in the liquid, but not the vapor phase. Furthermore, it is convenient to assume that the change in composition is brought about by a change $dn_2^{(l)}$ in the number of moles of solute in the liquid; $n_1^{(l)}$, the number of moles of solvent, remaining fixed. Under these conditions, we may use Equation (9.10) for the solvent in the vapor phase

$$d\mu_1^{(v)} = v^{(v)0}\,dp - s^{(v)0}\,dT, \tag{9.24}$$

and for the liquid phase Equation (9.9) becomes

$$d\mu_1^{(l)} = \bar{v}_1^{(l)}\,dp - \bar{s}_1^{(l)}\,dT + \left[\frac{\partial \mu_1^{(l)}}{\partial n_2^{(l)}}\right]_{T,p} dn_2^{(l)}. \tag{9.25}$$

If it is assumed that the liquid phase is an ideal solution, Equation (9.1) may be used in the derivative on the right of Equation (9.25) so that

$$d\mu_1^{(l)} = \bar{v}_1^{(l)}\,dp - \bar{s}_1^{(l)}\,dT - RT\left\{\frac{1}{n_1^{(l)} + n_2^{(l)}}\right\} dn_2^{(l)}. \tag{9.26}$$

Substitution of Equations (9.24) and (9.26) into Equation (9.8), and setting $dp = 0$, since the environmental pressure is fixed, yields

$$[s_1^{(v)0} - \bar{s}_1^{(l)}]\,dT = RT\left\{\frac{1}{n_1^{(l)} + n_2^{(l)}}\right\} dn_2^{(l)}. \tag{9.27}$$

Now the entropy difference in brackets on the left can be shown as in the case of Equation (9.15) to be $\frac{\lambda_{1(v)}}{T}$ where $\lambda_{1(v)}$ is the heat of vaporization of component 1 at constant T and p. However, in this case, care must be taken to emphasize that constant composition is also implied. Thus, $\lambda_{1(v)}$ may be regarded as the heat absorbed reversibly by the system when one mole of species 1 is transferred from the liquid to the vapor phase in a system of infinite extent. With this identification, Equation (9.27) becomes

$$dT_B = \frac{RT_B^2}{\lambda_{1(v)}}\left\{\frac{1}{n_1^{(l)} + n_2^{(l)}}\right\} dn_2^{(l)}, \tag{9.28}$$

where we have appended the subscript B to T to emphasize that the variation in boiling point is under consideration. Equation (9.28) determines the manner in which the boiling point changes with a change in solute content $n_2^{(l)}$.

If attention is restricted to dilute solutions,

$$n_1^{(l)} \gg n_2^{(l)}, \tag{9.29}$$

and Equation (9.28) may be approximated by

$$dT_B = \frac{RT_B^2\,dn_2^{(l)}}{\lambda_{1(v)}n_1^{(l)}}. \tag{9.30}$$

To a high degree of approximation, the differentials may be replaced by increments

$$\Delta T_B = \frac{RT_B^2\,\Delta n_2^{(l)}}{\lambda_{1(v)}n_1^{(l)}} = \frac{RT_B^2 n_2^{(l)}}{\lambda_{1(v)}n_1^{(l)}}, \tag{9.31}$$

where, in view of the diluteness of the solution, the increment $\Delta n_2^{(l)}$ may be identified with the entire solute content $n_2^{(l)}$. Furthermore, $\lambda_{1(v)}$ may be taken as the heat of vaporization of pure solvent. Since $\lambda_{(1)v}$ is positive, ΔT_B represents a boiling point *elevation*.

This expression is frequently employed in the determination of molecular weights. For example, if ω_2 is the mass in grams of species 2 in the solution, and ω_1 the mass of species 1, then

$$n_1^{(l)} = \frac{\omega_1}{m_1}, \qquad n_2^{(l)} = \frac{\omega_2}{m_2}, \tag{9.32}$$

where m_1 and m_2 are the molar weights of the respective species. Substitution of Equation (9.32) into Equation (9.31) gives

$$m_2 = \frac{RT_B^2 \omega_2 m_1}{\Delta T_B \lambda_{1(v)} \omega_1}, \tag{9.33}$$

so that if all the quantities on the right are known, m_2 may be determined.

Similar equations may be derived for the freezing point phenomenon. Here, if the solid consists of a single component (like the vapor phase in the case just considered), that is, if the solute is confined to the liquid, the freezing point of the solvent is *depressed*, provided that the heat of fusion is positive. Freezing point depressions may also be used in molecular weight determinations. The appropriate formula has T_f, the freezing point, in place of T_B in Equation (9.33), and $\lambda_{1(f)}$, the heat of fusion, in place of $\lambda_{1(v)}$.

When both phases are accessible to the solute, the situation is more complicated; but, in general, it may be dealt with in the same manner.

5. Chemical Equilibrium and the Law of Mass Action

The laws of ideal solution may be used also to facilitate the study of chemical equilibrium. Consider an ideal solution whose components may undergo chemical reaction. The chemical process may be symbolized by the relation

$$\sum_i \nu_i A_i = 0. \tag{9.34}$$

Here A_i represents the stoichiometric formula for the ith species. Thus, for water A_i is H_2O. The quantity ν_i is the stoichiometric coefficient, that is, the number of molecules of water taking part in the balanced reaction. If a species i does not participate in the reaction, for example if it is an inert solvent, then $\nu_i = 0$. If species i is a product or resultant of the reaction, ν_i is positive. If it is a reactant, ν_i is negative.

Usually the chemical phase within which the reaction is proceeding is constrained so that it may only perform volume work. If this is so, then the Gibbs free energy G may be employed as a thermodynamic potential and the requisite extremal condition is (see Equation (8.32)):

$$(\delta G)_{T,p} = 0. \tag{9.35}$$

Substitution of Equation (8.56) without the superscripts into Equation (9.34) yields

$$\sum_i \mu_i \, \delta n_i = 0. \tag{9.36}$$

If the variations δn_i are entirely due to the chemical reaction, it follows that

$$\frac{\delta n_i}{\delta n_k} = \frac{\nu_i}{\nu_k}, \qquad i,k = 1, 2, \ldots, C, \tag{9.37}$$

and substitution of Equation (9.37) into Equation (9.36) then yields

$$\sum_i \nu_i \mu_i = 0. \tag{9.38}$$

If the solution is ideal so that μ_i may be replaced by the right side of Equation (9.1), condition (9.38) becomes

$$\prod_i x_i^{\nu_i} = \exp\left\{-\frac{\sum_i \nu_i \mu_i^0}{RT}\right\} = K^*(T,p). \tag{9.39}$$

It is clear that the quantity K^* depends only upon T and p and not upon composition, since this is true of the various μ_i^0. $K^*(T,p)$ is referred to as the *equilibrium constant*, and Equation (9.39) is the familiar law of mass action. The quantity

$$\sum_i \nu_i \mu_i^0 \tag{9.40}$$

is the change in Gibbs free energy which would attend the occurrence of a unit of reaction (could it take place) among the reactants in the pure state. If the pure states are regarded as *standard* states, the various μ_i^0 may be referred to as the chemical potentials in the standard state, and expression (9.40) may be referred to as the free energy of reaction in the standard state.

In dilute solution (with an inert solvent), the various x_i are proportional to concentrations c_i. Under these conditions Equation (9.39) may be expressed as

$$\prod_i c_i^{\nu_i} = K_c^*(T,p), \tag{9.41}$$

in which K_c^* rather than K^* is used because the dimensions of the equilibrium constant have been changed. This is the most common form of the law of mass action.

6. Thermodynamic Activity, Dilute Solutions

Equation (9.1) has been very useful, and it is unfortunate that its validity should be restricted to ideal solutions. If we attempted to represent the chemical potential of some component i of a nonideal solution by the form

(9.1), we would have to add to $\mu_i^0(T,p)$ a correction term $g_i(T,p,x_1, \ldots, x_{C-1})$ which depends upon composition as well as upon T and p. Thus, for a nonideal solution

$$\mu_i = \mu_i^0(T,p) + g_i(T,p,x_i, \ldots, x_{C-1}) + RT \ln x_i. \tag{9.42}$$

The appearance of the correction term g_i may be softened by writing

$$g_i = RT \ln f_i, \tag{9.43}$$

where $f_i(T,p,x_1, \ldots, x_{C-1})$ is called the *activity coefficient*. Introduction of Equation (9.43) into Equation (9.42) yields

$$\mu_i = \mu_i^0(T,p) + RT \ln f_i x_i. \tag{9.44}$$

The quantity $f_i x_i$ is usually referred to as the *thermodynamic activity* of species i and is represented by the symbol a_i. Thus,

$$a_i = f_i x_i. \tag{9.45}$$

In terms of Equation (9.45) we may write Equation (9.44) as

$$\mu_i = \mu_i^0(T,p) + RT \ln a_i. \tag{9.46}$$

This form resembles Equation (9.1) but this is entirely artificial, because the dependence of a_i upon composition can only be determined by experiment, or through the use of theoretical techniques which go beyond thermodynamics. In this sense, a_i is simply an adjustable parameter. In fact, most methods for determining a_i really depend upon an exact determination of μ_i (although sometimes μ_i may not appear explicitly in the formulas employed).[9]

On the other hand, the quantity f_i is a convenient index of the departure of a given solution from ideality. Furthermore, it may be shown from considerations of molecular theory, and even from certain plausibility arguments based on macroscopic reasoning,[10] that in dilute solutions the chemical potential of the solute may be represented by the form

$$\mu_i = D_i(T,p) + RT \ln x_i, \qquad x_i \rightarrow 0, \tag{9.47}$$

where D_i, although a function of T and p only, does not (like μ_i^0) usually represent the chemical potential of pure solute. Of course, if the solution is both *ideal* and *dilute*, D_i does represent the chemical potential of pure solute.

---

[9] K. S. Pitzer and L. Brewer, *Thermodynamics* (revision of the book by G. N. Lewis and M. Randall) (New York: McGraw-Hill Book Company, Inc., 1961), Chapter 20.
[10] *Ibid.*, Chapter 19.

Suppose $D_i(T,p)$ is added and subtracted on the right side of Equation (9.46). Then,

$$\mu_i = D_i + (\mu_i^0 - D_i) + RT \ln a_i$$
$$= D_i + RT \ln a_i \exp\left\{\frac{\mu_i^0 - D_i}{RT}\right\}$$
$$= D_i + RT \ln a_i' \tag{9.48}$$

where

$$a_i' = a_i \exp\left\{\frac{\mu_i^0 - D_i}{RT}\right\}. \tag{9.49}$$

Since a_i was an adjustable parameter to begin with, we may use a_i' as easily as a_i and *define* $D_i(T,p)$ as the chemical potential in the standard state. Since the exponential in Equation (9.49) does not depend upon composition, a_i' is proportional to a_i as long as temperature and pressure remain fixed.

It is clear from Equation (9.49) that the numerical value of the activity depends upon the choice of standard state, but that for different standard states the *relative* values of activity at different compositions remain the same. The choice of that a_i' which goes with the standard state function, D_i, is especially convenient because in dilute solution (see Equation (9.47))

$$a_i' \rightarrow x_i. \tag{9.50}$$

With the use of either Equation (9.46) or Equation (9.48), all the formulas derived in this chapter remain valid provided that x_i is replaced by a_i or a_i'. It must be remembered, however, that the activity is only an adjustable parameter possessing, at the most, a limiting identification, namely Equation (9.50), with the mole fraction. In many instances this limiting relation holds approximately even in solutions which are relatively concentrated.

In closing this section, it is fitting to indicate that once the form (9.47) has been assumed for the chemical potential of a solute in a dilute solution, it may be demonstrated by thermodynamic reasoning alone that Equation (9.1) holds for the *solvent*, that is, that the solvent behaves like a component of an *ideal* solution. To prove this, we make use of the Gibbs-Duhem relation (Equation (8.71)), assigning to the solvent the subscript 1. The crucial step here is the use of Equation (8.71) at constant temperature and pressure so as to remove $D_i(T,p)$ from the problem. Thus,

$$n_1\, d\mu_1 = -\sum_{i=2}^{i=C} n_i\, d\mu_i = -RT\sum_{i=2}^{i=C} n_i d \ln x_i, \tag{9.51}$$

where the last form results from the use of Equation (9.47) for each of the solutes. Now,

$$\sum_{i=2}^{i=C} n_i\, d \ln x_i = \sum_{i=2}^{i=C} n_i \frac{dx_i}{x_i} = \sum_{i=2}^{i=C} \frac{n_i\, dx_i}{\left(n_i \Big/ \sum_{k=1}^{k=C} n_k\right)} = \sum_{k=1}^{k=C} n_k \sum_{i=2}^{i=C} dx_i = -dx_1 \sum_{k=1}^{k=C} n_k. \tag{9.52}$$

Substitution of this into the last term in Equation (9.51) gives

$$d\mu_1 = RT\,d\ln x_1, \tag{9.53}$$

which upon integration yields

$$\mu_1 = \mu_1^0(T,p) + RT\ln x_1, \tag{9.54}$$

where μ_1^0 is the constant of integration. This is identical with Equation (9.1).

7. The Measurement of G, Electrochemical Cells

This chapter would be incomplete without some mention being made of the techniques available for the experimental determination of G. Since $G = U + pV - TS = H - TS$, there is always the possibility of employing Equations (6.33) and (6.34), when heat capacity data are available, to compute H and S, and therefore G, relative to the reference state (T_0,p_0). On the other hand, one is often interested in knowing the change, ΔG, which corresponds to a given process; and there are more direct methods available which do not require the detailed measurement of heat capacity.

As an example, we shall consider the use of electrochemical cells. This will serve the double purpose of permitting us to discuss, quantitatively, this system which we have previously employed in a qualitative manner in order to elaborate certain conceptual ideas, while at the same time permitting us to focus on a concrete example of $\Delta w_{1_{\text{rev}}}$ which played such an important role in Chapter 8. It will be convenient to limit consideration to the simplest of systems and to treat them schematically. Therefore, we consider the cell illustrated in Figure 9.1.

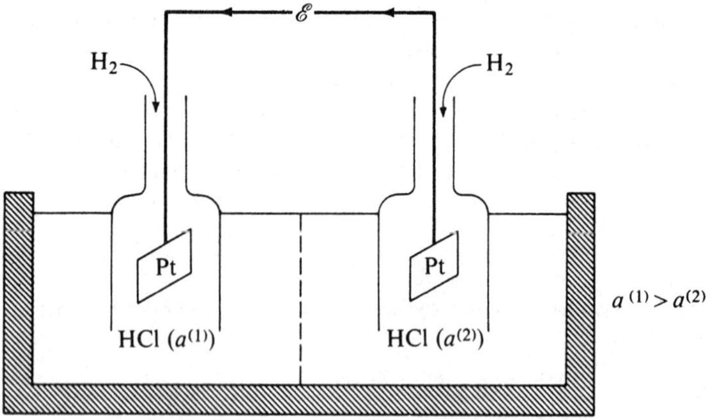

FIGURE 9.1. *Schematic representation of* HCl *concentration cell.*

This cell consists of two compartments, each containing an aqueous solution of hydrochloric acid, the solution on the left having HCl activity $a^{(1)}$, and the solution on the right having activity $a^{(2)}$. It is assumed that $a^{(1)}$ exceeds $a^{(2)}$ so that when the solutions are placed in contact with one another, HCl will tend to diffuse from left to right until the activities (and therefore the chemical potentials) have been equalized. The dashed vertical line in the center of the cell represents some physical device, such as a glass frit, designed to slow the diffusion process when the solutions are in contact.

The cell process may then be represented by a chemical equation

$$\mathrm{HCl}(a^{(1)}) \rightarrow \mathrm{HCl}(a^{(2)}). \tag{9.55}$$

If the diffusion process is slowed sufficiently, it is possible to provide an alternative path for the reaction (9.55). Thus, into each compartment of the cell a platinum electrode may be inserted as shown in Figure 9.1. In addition, a housing may be placed over each electrode through which gaseous hydrogen may be led into the solution at some fixed pressure p. The following series of events will now occur.

(It should be emphasized that the description of this series of events, taking place on the molecular level, represents an extra-thermodynamic point of view. We undertake the description merely to provide the reader with some insight into the mechanism; but the entire thermodynamic theory could be developed, with the exception of the "junction potential" without any reference to mechanism.)

The hydrochloric acid solutions are partially ionized so there are present in solution both H^+ and Cl^- ions. As a result of the presence of H_2 gas at the platinum surface, the following equilibrium exists:

$$\mathrm{H_2}\ (\text{gas}) \rightleftarrows 2\mathrm{H^+}\ (\text{solution}) + 2\mathrm{e^-}\ (\text{electrode}), \tag{9.56}$$

where e^- symbolizes an electron. Now reaction (9.55) may take place as follows. At the electrode on the right, the reaction (9.56) takes place in the forward direction so that more H^+ ions enter the solution and electrons enter the electrode. These electrons pass through the circuit connecting the electrodes in the direction indicated by the arrow in Figure 9.1, and upon arriving at the electrode on the left are consumed by the reaction (9.56), which now occurs in the reverse sense. This reverse reaction, of course, extracts hydrogen ions from solution and transforms them into a quantity of hydrogen gas equal to the amount consumed at the electrode on the right. The net effect, therefore, has been the transfer of an H^+ ion from the solution on the left to the one on the right. During this time, a Cl^- ion is transferred *through the solution* from left to right so as to maintain charge balance. The overall effect is therefore equivalent to the process of Equation (9.55).

Now the flow of current through the external circuit implies the existence of an emf $\mathscr{E}$. This can be measured by suppressing the current by the application of a counter-emf. We have glossed over certain important details. For example, the cell may be called a *concentration cell* since it places two solutions of the same species at different concentrations in contact with one another. At the junction between the two solutions there will be a so-called *junction potential* which is a nonthermodynamic and irreversible quantity for which the measurement of the cell emf must be corrected. This important limitation will be ignored in the present discussion.

The cell reaction (9.55) can be made to occur reversibly by the application of an external emf which in the present instance provides an example of a variational constraint χ_1. In such a reversible process the system may perform volume work (since a change in volume accompanies the reaction) as well as electrical work against the applied emf. This electrical work may be computed, using Equation (1.3), by setting

$$dQ = \mathscr{F}\, dn, \tag{9.57}$$

where $\mathscr{F}$ is the number of coulombs of electrons which flow through the external circuit to accompany one mole of reaction. $\mathscr{F}$ is called the Faraday and is equal to 96,487 coulombs. In Equation (9.57) dn stands for dn moles of reaction. Thus, the electrical work performed by the cell during dn moles of reaction will be, according to Equations (9.57) and (1.3),

$$Dw = \mathscr{E}\mathscr{F}\, dn. \tag{9.58}$$

If the reaction takes place at constant temperature and pressure, then the system is following a constant T, p reversible path, tending to reduce $\mathscr{E}$ to zero, that is, tending to remove the constraint χ_1. Thus, this is a process inverse to the one to which Equation (8.12) refers. As a result, Δw in Equation (9.58) is the negative of $-\Delta w_{1_{\text{rev}}}$ in Equation (8.12) so that

$$\Delta w = +\Delta w_{1_{\text{rev}}} = \mathscr{F}\int \mathscr{E}\, dn, \tag{9.59}$$

where dn still refers to the process in which the system goes from the *end* state to the *initial* state in the parlance of the discussion surrounding Equation (8.12). If we wish to have dn positive for the process, going from the initial to the end state, then $-dn$ must be substituted for $+dn$ in Equation (9.59). The result is

$$\Delta w_{1_{\text{rev}}} = -\mathscr{F}\int \mathscr{E}\, dn, \tag{9.60}$$

where $\Delta w_{1_{\text{rev}}}$ is the same quantity appearing in Equation (8.12), and dn is now positive when the inverse of reaction (9.55) takes place.

Since the path of the reversible process is one of constant temperature and pressure, and the only other kind of work, besides electrical, is volume work, Equation (8.28) demands that $\Delta\varphi$, and therefore (according to Equation (8.12)) $-\Delta w_{1_{rev}}$, measure ΔG. Then Equation (9.60) requires

$$\Delta G = \mathscr{F} \int \mathscr{E}\, dn, \tag{9.61}$$

where Δ refers to the process inverse to the one defined by Equation (9.55). The free energy change per mole of reaction is then

$$\left[\frac{\partial \Delta G}{\partial n}\right]_{T,p} = \mathscr{E}\mathscr{F} \tag{9.62}$$

and is obviously measurable by the straightforward measurement of $\mathscr{E}$, the cell emf.

It is interesting to note, in connection with Equation (8.12), that in this section the details of the *end state* produced by the variational constraint *are* of explicit interest, unlike the situation confronted in the discussion of thermodynamic potentials, where the explicit nature of the end state was not important. In fact, the cell reaction proceeds between two easily observable states, one of which *is* the *end state.*

Since the chemical potential is the partial molar Gibbs free energy, it follows, for the reverse of process (9.55), that for a mole of reaction the change in free energy ($\mathscr{E}\mathscr{F}$ by Equation (9.62)) is given by

$$\mathscr{E}\mathscr{F} = \mu^{(1)} - \mu^{(2)}. \tag{9.63}$$

If the activities of HCl are $a^{(1)}$ and $a^{(2)}$, respectively, then use of Equation (9.46) in Equation (9.63) yields

$$\mathscr{E}\mathscr{F} = RT \ln \frac{a^{(1)}}{a^{(2)}}, \tag{9.64}$$

so that a measurement of $\mathscr{E}$ makes possible the determination of the activity ratio. Equations (9.63) and (9.64) represent one of the methods by means of which relative chemical potentials and relative activities may be measured.

X

Osmotic Systems

1. More on the Generalized Feature of Equilibrium

It has been suggested in Chapter 8, Section 6, that the treatment of systems which are constrained very generally (and are therefore capable of other kinds of work besides volume work) should proceed with Equation (8.33) as a foundation.

The development of suitable thermodynamic potentials should properly be based on the finite incremental expression (8.12). Equation (8.33) is merely the differential version of Equation (8.12) and is more convenient mathematically although, in some cases, more restrictive than (8.12). From the purely formal point of view, no error is committed by working with the differential expression in developing thermodynamic potentials since the differential terms in Equation (8.33) are in one-to-one correspondence with the incremental terms in Equation (8.12).

Furthermore, if we are interested in "features" of equilibrium, that is, extremal conditions like Equations (8.29) through (8.32), the eventual goal *is* a differential expression and so the utilization of Equation (8.33), itself a differential expression, represents the most direct route. We shall adopt this point of view in the next few chapters and seek thermodynamic potentials by looking for functions of state whose differentials, over selected paths in state space, *imitate* $D\varphi$. This, of course, is the same procedure which was utilized in Chapter 8, Section 5, where functions were sought whose *increments* over selected paths *imitated* $\Delta\varphi$.

Whenever a new kind of thermodynamic system (capable of new kinds of work) is involved, one may minimize confusion at the outset by starting with Equation (8.33). The fundamental problem resides in the correct specification of $Dw_{0_{\text{rev}}}$, which of course depends upon the kind of work which the system may perform against the initial constraints χ_0. When this work is properly specified, functions of state may be devised (as in Chapter 8) whose differentials *imitate* $D\varphi$.

It is somewhat unfortunate that most texts on thermodynamics tend to label all thermodynamic potentials as free energies. This is accomplished by simply adding terms symmetrical in other work variables to the pV which appears in the Gibbs free energy. It is as if one wished to prove the validity of the new potential, arguing from analogy, rather than by a return to fundamentals. This is especially unfortunate when it draws attention away from general fundamental expressions like Equation (8.33) which are designed to emphasize the *differences* as well as the similarities between thermodynamic systems. This course will not be followed here, but instead, wherever possible, an attempt will be made to show how a thermodynamic potential differs from the familiar Gibbs free energy.

In each development we shall return again and again to Equation (8.33) until the reader sees how this important expression may be made the basis for the treatment of *any* system. In this way an essential unity will be provided which embraces all thermodynamic argument, and in the light of which the domain of application of thermodynamics will be more clearly delineated. It is hoped that the study of this chapter and the few that follow will enable the reader to employ thermodynamics, in any relevant circumstance, without having to draw either upon analogy or upon procedures learned by rote.

In the remainder of this chapter osmotic systems will be treated by the method of thermodynamic potentials. The treatment will not be exhaustive. In keeping with the purpose of this book, attention will be focused on methods rather than on details. The reader who is interested in a more thorough analysis is advised to consult other texts, some of which have been recommended in the Preface.

2. Osmotic Systems

The osmotic system is the simplest in which the thermodynamic potential is not a free energy in the elementary sense defined by Equation (4.71). This departure from form is connected directly with the appearance of more than one term in $Dw_{0_{\text{rev}}}$. Although the several terms are all still *volume* works, these are performed at distinct pressures and are separately identifiable. It is in this sense that we are confronted with different *kinds* of work.

This point will be illustrated with a simple closed system consisting of two phases; one (phase 1) is composed of a single component and the other (phase 2) is composed of two components. As usual, phases will be denoted by superscripts, while the components will be designated by subscripts. The osmotic character of the system is provided by a rigid semipermeable membrane which separates phase 1 from phase 2, and which passes component 1 (which we shall call the solvent) but not component 2 (which may be called the solute). The system is diagrammed in Figure 10.1. Since the membrane is rigid, different pressures may be applied to the phases by means

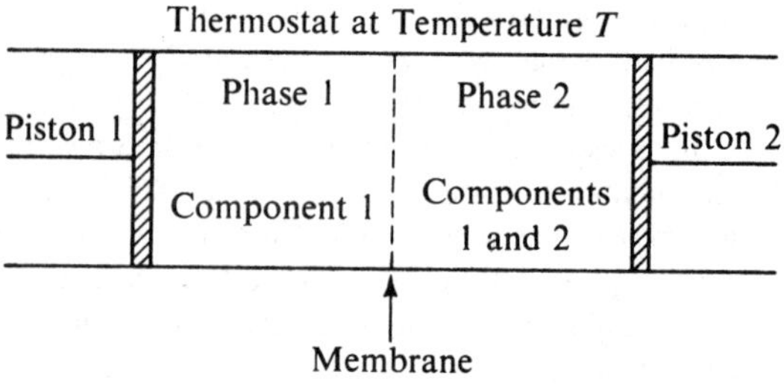

FIGURE 10.1. *Example of Osmotic Experiment.*

of pistons 1 and 2. Thus, the total reversible work $Dw_{0_{\text{rev}}}$ performed in connection with the constraints χ_0 (which in the present case are the pistons 1 and 2) is

$$Dw_{0_{\text{rev}}} = p^{(1)}\,dV^{(1)} + p^{(2)}\,dV^{(2)}, \tag{10.1}$$

and for this system Equation (8.33) becomes

$$D\varphi = \delta U - T\,\delta S + p^{(1)}\,\delta V^{(1)} + p^{(2)}\,\delta V^{(2)} = 0. \tag{10.2}$$

The procedure now, as always, involves the invention and synthesis of functions of state whose differentials *imitate* $D\varphi$ over selected paths in state space (paths over which the system may be driven by the application of additional constraints).

For the problem under consideration (just as in the case of the system in Chapter 8), a number of useful thermodynamic potentials may be constructed in this manner. Since these are logically equivalent, the argument may be advanced with but one, and for conciseness we shall only deal with one. However, the reader may wish to try his hand at the construction of others.

Consider the following function:

$$\phi = U + p^{(1)}V^{(1)} + p^{(2)}V^{(2)} - TS, \tag{10.3}$$

Clearly ϕ is a function of state since it is composed of functions of state. It resembles the function G defined in Equation (4.71); however, it is not a Gibbs free energy in the elementary sense since it contains *two* distinct pV terms rather than *one*. Again, it must be emphasized that this point of view is not merely one of semantic convenience but is fundamental to the understanding of the problem in its relation to Equation (8.33). To call ϕ a Gibbs free energy opens the door to "understanding-by-analogy," a method which we wish to avoid. If a path in state space is selected along which $p^{(1)}$, $p^{(2)}$, and T are maintained constant, then from Equation (10.2)

$$[D\varphi]_{T,p^{(1)},p^{(2)}} = \delta U - T\,\delta S + p^{(1)}\,\delta V^{(1)} + p^{(2)}\,\delta V^{(2)}, \tag{10.4}$$

and from Equation (10.3)

$$[\delta\phi]_{T,p^{(1)},p^{(2)}} = \delta U - T\,\delta S + p^{(1)}\,\delta V^{(1)} + p^{(2)}\,\delta V^{(2)}. \tag{10.5}$$

Comparison of Equation (10.5) with Equation (10.4) shows that $d\phi$ *imitates* $D\varphi$ along a path of constant T, $p^{(1)}$, and $p^{(2)}$. Thus, ϕ is a thermodynamic potential for this system, and we may write

$$[\delta\phi]_{T,p^{(1)},p^{(2)}} = 0 \tag{10.6}$$

as the extremal condition. Temperature is maintained uniform and the same in both phases by the thermostat (see Figure 10.1) surrounding the system. As usual, the total volume V and entropy S may be decomposed into parts belonging to each phase. Thus

$$U = U^{(1)} + U^{(2)} \tag{10.7}$$

and

$$S = S^{(1)} + S^{(2)}, \tag{10.8}$$

from which it follows, by insertion of these relationships in Equation (10.3), that

$$\begin{aligned}\phi = [U^{(1)} + p^{(1)}V^{(1)} - TS^{(1)}]& \\ + [U^{(2)} + p^{(2)}V^{(2)} - TS^{(2)}] &= G^{(1)} + G^{(2)}.\end{aligned} \tag{10.9}$$

The Gibbs free energy enters the scene, not as a thermodynamic potential, but only because the thermodynamic potential is the sum of two Gibbs functions, one for each phase.

Introduction of Equation (10.9) into Equation (10.6) yields

$$\begin{aligned}[\delta\phi]_{T,p^{(1)},p^{(2)}} &= [\delta G^{(1)}]_{T,p^{(1)}} + [\delta G^{(2)}]_{T,p^{(1)}} \\ &= [V^{(1)}\,\delta p^{(1)} - S^{(1)}\,\delta T + \mu_1^{(1)}\,\delta n_1^{(1)}]_{T\,p^{(1)}} \\ &\quad + [V^{(2)}\,\delta p^{(2)} - S^{(2)}\,\delta T + \mu_1^{(2)}\,\delta n_1^{(2)} + \mu_2^{(2)}\,\delta n_2^{(2)}]_{T,p^{(2)}} \\ &= \mu_1^{(1)}\,\delta n_1^{(1)} + \mu_1^{(2)}\,\delta n_1^{(2)} + \mu_2^{(2)}\,\delta n_2^{(2)} = 0,\end{aligned} \tag{10.10}$$

where Equation (8.56) has been used, together with the condition that $n_2^{(1)} = 0$. Since the system is closed,

$$\delta n_1^{(2)} = -\delta n_1^{(1)}, \tag{10.11}$$

and because it is closed and the membrane is impermeable to component 2,

$$\delta n_2^{(2)} = 0. \tag{10.12}$$

Substitution of Equations (10.11) and (10.12) into Equation (10.10) gives

$$[\mu_1^{(1)} - \mu_1^{(2)}]\,\delta n_1^{(1)} = 0$$

or

$$\mu_1^{(1)} = \mu_1^{(2)}. \tag{10.13}$$

Thus, even though the pressure may be different in each phase, the maintenance of internal equilibrium still requires that the chemical potential of exchangeable component be the same in each phase.

If the pressures are different, the compositions of the two phases will be different. This is the essence of osmotic equilibrium. The difference in pressures

$$p^{(2)} - p^{(1)} = \pi \tag{10.14}$$

is called the osmotic pressure, and it is of interest to see how π is related to the composition of phase 2. To acquire insight into this question, it is necessary to know $\mu_1^{(2)}$ as a function of T, $p^{(2)}$, $n_1^{(2)}$, and $n_2^{(2)}$.

Equation (9.9) may be applied to component 1 (the solvent) in each phase. Thus,

$$d\mu_1^{(1)} = V_1^{(1)0}\,dp^{(1)} - S_1^{(1)0}\,dT, \tag{10.15}$$

$$d\mu_1^{(2)} = \bar{V}_1^{(2)}\,dp^{(2)} - \bar{S}_1^{(2)}\,dT + \left[\frac{\partial\mu_1^{(2)}}{\partial n_1^{(2)}}\right]_{T,p^{(2)}} dn_1^{(2)} + \left[\frac{\partial\mu_1^{(2)}}{\partial n_2^{(2)}}\right]_{T,p^{(2)}} dn_2^{(2)}. \tag{10.16}$$

In Equation (10.15) the subscript has been retained even though the phase consists of a single component so that the species may be identified with the solvent in phase 2. Since Equation (10.13) holds in the osmotic system, we must apply Equation (9.8) in order to maintain equilibrium in the face of changes in the values of the independent variables.

An interesting and convenient set of changes in these variables is the following. Assume that both phases consist initially of pure solvent, both at the pressure $p^{(1)}$, so that π defined by Equation (10.14) is initially zero. Then holding $n_1^{(1)}$ and $n_1^{(2)}$ fixed, $n_2^{(2)}$ moles of solute are added to phase 2, while the pressure in phase 1 is kept at $p^{(1)}$. Since T is constant, the only variable which may be adjusted to counter-balance the effect of varying $n_2^{(2)}$ is $p^{(2)}$. Furthermore, since neither temperature nor pressure is changed in phase 1, $d\mu_1^{(1)} = 0$. But then Equation (9.8) requires that $d\mu_1^{(2)}$ also be zero for the maintenance of equilibrium.

By introducing this requirement that $d\mu_1^{(2)}$ be zero into Equation (10.16), along with the conditions stated previously that $dT = 0$ and $dn_1^{(1)} = 0$, we obtain

$$\bar{V}_1^{(2)}\,dp^{(2)} = -\left[\frac{\partial\mu_1^{(2)}}{\partial n_2^{(2)}}\right]_{T,p^{(2)}} dn_2^{(2)}. \tag{10.17}$$

This equation may be integrated subject to the initial condition

$$p^{(2)} = p^{(1)}, \qquad n_2^{(2)} = 0. \tag{10.18}$$

The result is

$$p^{(2)} - p^{(1)} = \pi = -\int_0^{n_2^{(2)}} \frac{1}{\bar{V}_1^{(2)}}\left[\frac{\partial\mu_1^{(2)}}{\partial n_2^{(2)}}\right]_{T,p} dn_2^{(2)}, \tag{10.19}$$

Equation (10.19) relates the osmotic pressure to $n_2^{(2)}$, the amount of solute in phase 2. To obtain a more explicit relation, the dependence of $\mu_1^{(2)}$ on composition must be known, and as we have indicated previously, this requires extra-thermodynamic information.

In this connection, some insight can be gained by assuming that phase 2 is ideal. If we do this and employ Equation (9.1),

$$\begin{aligned}\mu_1^{(2)} &= \mu_1^{(2)0}(T,p^{(2)}) + RT\ln x_2^{(2)} \\ &= \mu_1^{(2)0}(T,p^{(2)}) + RT\ln n_1^{(2)} - RT\ln \{n_1^{(2)} + n_2^{(2)}\},\end{aligned} \qquad (10.20)$$

and

$$\left[\frac{\partial \mu_1^{(2)}}{\partial n_2^{(2)}}\right]_{T,p^{(2)}} = -\frac{RT}{\{n_1^{(2)} + n_2^{(2)}\}}. \qquad (10.21)$$

Furthermore, according to Equation (5.70),

$$\bar{V}_1^{(2)} = V_1^{(2)0} = V_1^0, \qquad (10.22)$$

where V_1^0 is the volume per mole of the pure solvent at temperature T and pressure $p^{(2)}$. Substitution of Equations (10.21) and (10.22) into Equation (10.19) yields

$$\pi = \frac{RT}{V_1^0}\int_0^{n_2^{(2)}} \frac{dn_2^{(2)}}{\{n_1^{(2)} + n_2^{(2)}\}} = -\frac{RT}{V_1^0}\ln x_1^{(2)}, \qquad (10.23)$$

(where the small dependence of V_1^0 on $p^{(2)}$ has been ignored) which shows how the osmotic pressure depends upon the composition of an ideal binary (two-component) solution.

If the solution is dilute with respect to the solute,

$$x_2^{(2)} = 1 - x_1^{(2)} \ll 1. \qquad (10.24)$$

By substituting this into Equation (10.23) we get

$$\pi = -\frac{RT}{V_1^0}\ln(1 - x_2^{(2)}) \approx \frac{RT}{V_1^0}x_2^{(2)} = \frac{RT}{V_1^0}\frac{n_2^{(2)}}{\{n_1^{(2)} + n_2^{(2)}\}}. \qquad (10.25)$$

Since the solution is dilute,

$$n_1^{(2)} + n_2^{(2)} \approx n_1^{(2)}. \qquad (10.26)$$

Furthermore, since the solution is dilute and V_1^0 is the volume per mole of pure solvent,

$$n_1^{(2)}V_1^0 \approx V^{(2)}, \qquad (10.27)$$

where $V^{(2)}$ is the volume of phase 2. Introduction of Equations (10.26) and (10.27) into Equation (10.25) yields

$$\pi V^{(2)} = RTn_2^{(2)}. \qquad (10.28)$$

This equation is formally similar to the ideal gas law (5.1), the osmotic pressure π playing the role of the pressure p. Thus, in a dilute ideal solution we may regard the osmotic pressure formally as an "extra" pressure produced by $n_2^{(2)}$ solute molecules, behaving as an ideal gas in the volume $V^{(2)}$ of the solution.

XI

Systems Which May Perform Surface Work

1. Surface Layers

Reference has already been made (Chapter 1, Section 4) to systems which are so finely divided that surface quantities must be considered as independent variables of state. The thermodynamic treatment of such systems has a long history, beginning with Gibbs[11] and culminating (at least temporarily) in the work of Tolman,[12] Buff,[13] Cahn and Hillard,[14] and Hart.[15] This field constitutes an excellent example of the confusion which can be generated when one endeavors to treat new thermodynamic systems by "analogy" rather than by a return to the fundamental principles embodied in Equation (8.33).

As usual, the experimenter, on the basis of measurements performed on the system, must decide what new variables are to be included. In the treatment of surfaces, several stages of refinement are possible. In the first stage, one may view a surface (say the interface between a liquid and its vapor) as a two-dimensional membrane having no thickness, but possessing lateral cohesion, so that a tensile stress—the surface tension—develops when the membrane is extended. In this view the liquid and its vapor are separated by a mathematical surface of discontinuity at which the density of the liquid undergoes an abrupt transition to that of the vapor.

In a second stage of refinement, one assumes that the effects of the surface may be accounted for in terms of the surface area and surface tension, but recognizes that the transition in density between liquid and vapor is not perfectly abrupt. Instead, there exists a transition layer between the liquid

---

[11] J. W. Gibbs, *Collected Works* (New York: Longmans Green & Co., 1928), pp. 55–353.
[12] R. C. Tolman, *J. Chem. Phys.*, 16 (1948), 758 also, 17, pp. 118 and 333 (1949).
[13] F. P. Buff, *J. Chem. Phys.*, 19 (1951), 1591.
[14] J. W. Cahn and J. E. Hilliard, *J. Chem. Phys.*, 28 (1958), 258.
[15] E. W. Hart, *Phys. Rev.*, 113 (1959), 412; also 114 (1959), 27.

and vapor which may, in typical cases, be 10–15 angstroms thick, and in which the density changes continuously from that characteristic of the liquid to that of the vapor. As a result, problems arise (of the sort discussed in Chapter 3, Section 7) in the physical distinctness and mechanical separability of the two phases, and one cannot perform an accurate thermodynamic analysis without consideration of the structure of the surface layer.

In a third stage of refinement, it becomes necessary to consider how the curvature of a given surface affects surface tension[(12,13)], and curvature must be treated as an *additional* thermodynamic variable. However, investigations of this problem have revealed that the curvature must be exceedingly high (for example, one must consider states of subdivision such that individual particles are almost as small as molecules) before it has an appreciable effect.

In recent years an entirely different approach has been introduced into the field of surface thermodynamics.[(14,15)] In this approach, surface tensions and surfaces of separation (interfaces between phases) do not appear. Instead, quantities like the gradient of density in the transition layer and various of its derivatives are treated as thermodynamic variables. This technique appears to have several advantages, but we shall not discuss it here in view of our limited objectives.

In the next section, remarks will be confined to the first stage of refinement in which the surface is treated as an elastic membrane of zero thickness. This limitation will allow us to expose subtleties in the method without obscuring the essentials. In later sections, attention will be given to the second stage of refinement in which the structure of the transition layer must be considered.

2. The Laplace Relation

If the generalized feature of equilibrium (8.33) is to be used, we must express the surface contribution to $Dw_{0_{\text{rev}}}$ in an explicit form. If the surface tension is denoted by σ, then the reversible work which must be performed *on* the system in order to increase the surface area Ω by $d\Omega$ will be

$$\sigma \, d\Omega. \tag{11.1}$$

In effect, this is the thermodynamic definition of surface tension.

The relatively simple example consisting of a single component liquid drop surrounded by its own vapor is excellent for demonstrating the utility of Equation (8.33) in the uniform treatment of arbitrary thermodynamic systems. It is also valuable for demonstrating how sensitive $Dw_{0_{\text{rev}}}$ is to the exact configuration of the constraints $\mathbf{X}_0$. Consider the arrangement of Figure 11.1. Here we have a drop separated from its vapor by a surface of

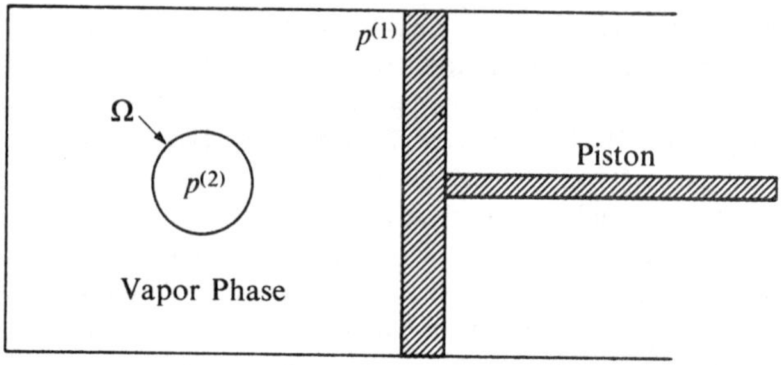

FIGURE 11.1. *Drop (phase 2) surrounded by its own vapor (phase 1) and making contact with the environment through the vapor.*

area Ω. The entire system is enclosed in a cylinder and may exchange work with the environment through the agency of the piston. The pressure in the drop is denoted by $p^{(2)}$ and may be different from that of the vapor, denoted by $p^{(1)}$. Notice that the piston makes contact with the system only through the vapor. Thus, in this configuration any work exchanged between the system and its environment will be volume work performed at the pressure $p^{(1)}$.

Now consider another configuration, Figure 11.2. This is the same as Figure 11.1 except that here direct contact is made with the drop by two "squeezers" which pass through pressure-tight seals in the sides of the cylinder. Now the system not only contacts the environment through the vapor phase at pressure $p^{(1)}$ but also through this direct connection with the drop. The work exchanged with the environment will now involve *surface*

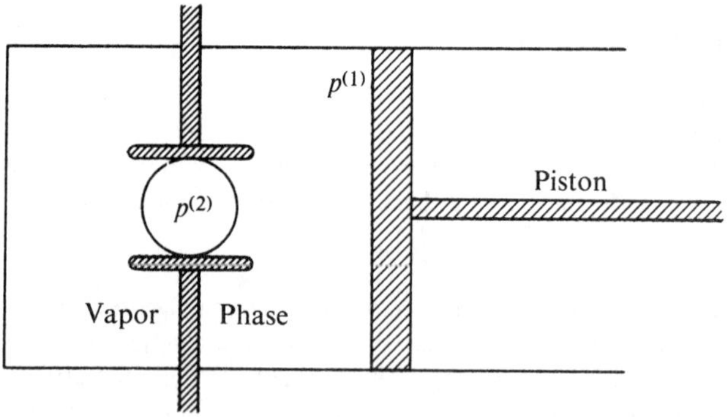

FIGURE 11.2. *System of Figure 11.1, this time constrained so that the drop makes direct contact with environment.*

work associated with the drop as well as the *volume* work, since squeezing will produce a liquid-"squeezer" interface in addition to the existing liquid-vapor interface.

Thus, the addition of another constraint (the "squeezers") makes possible the performance of additional work.

We shall concentrate on the system of Figure 11.1. Since it may only perform volume work at the pressure $p^{(1)}$,

$$Dw_{0_{\text{rev}}} = p^{(1)}\,dV, \tag{11.2}$$

where

$$V = V^{(1)} + V^{(2)} \quad \text{and} \quad dV = dV^{(1)} + dV^{(2)}. \tag{11.3}$$

V is the total volume and $V^{(1)}$ and $V^{(2)}$ are the volumes of the vapor and drop, respectively. Although the surface could be treated as a distinct entity, we shall, for convenience, regard it as belonging to the drop. Since the drop performs work on its own surface, the net work it performs on the vapor phase is

$$p^{(2)}\,dV^{(2)} - \sigma\,d\Omega, \tag{11.4}$$

while the vapor phase does the work

$$p^{(1)}\,dV^{(1)} \tag{11.5}$$

on *its* total surroundings (which include the drop) while it performs the work

$$-p^{(2)}\,dV^{(2)} + \sigma\,d\Omega \tag{11.6}$$

on the drop—obviously the negative of Equation (11.4). Thus, $Dw_{0_{\text{rev}}}$, communicated to the surroundings of the entire system, as a whole is also expressible as

$$\begin{aligned} Dw_{0_{\text{rev}}} &= p^{(1)}\,dV^{(1)} - \{-p^{(2)}\,dV^{(2)} + \sigma\,d\Omega\} \\ &= p^{(1)}\,dV^{(1)} + p^{(2)}\,dV^{(2)} - \sigma\,d\Omega. \end{aligned} \tag{11.7}$$

By substituting Equation (11.3) into Equation (11.2) and equating the right side of the resulting equation to the right side of Equation (11.7) we get

$$(p^{(2)} - p^{(1)})\,dV^{(2)} = \sigma\,d\Omega. \tag{11.8}$$

The connection between Ω and $V^{(2)}$ is purely geometric.† Thus, if the radius of the drop is r, it follows that

$$\begin{aligned} \Omega &= 4\pi r^2, \\ V^{(2)} &= \tfrac{4}{3}\,\pi r^3, \end{aligned} \tag{11.9}$$

† Notice that the condition that the drop remain spherical is a constraint which we apply to the virtual variations as well as the initial state. However, it can be shown that in the absence of any constraint forbidding the spherical form, the equilibrium shape in the initial state will be spherical. We shall not pursue this argument here.

so that

$$d\Omega = 8\pi r\, dr,$$
$$dV^{(2)} = 4\pi r^2\, dr. \tag{11.10}$$

By introducing Equation (11.10) into Equation (11.8) we get

$$p^{(2)} - p^{(1)} = \frac{2\sigma}{r}. \tag{11.11}$$

This is the famous Laplace relation between the pressures inside and outside of a spherical drop. It is a purely mechanical (nonthermal) relation, since its derivation is based upon arguments which involve the concept of work only. One can, of course, derive it by methods other than the one used here.

3. The Thermodynamic Potential

To simplify our considerations further, assume that the drop consists of an incompressible liquid and that σ depends only upon temperature. (This is an extra-thermodynamic assumption and represents a specialization of the system.)

Now substitute Equation (11.2) with Equation (11.3) into Equation (8.33). The result is

$$D\varphi = \delta U + p^{(1)}\,\delta V^{(1)} + p^{(1)}\,\delta V^{(2)} - T\,\delta S = 0. \tag{11.12}$$

This expression is similar to Equation (10.2) which arose in connection with the osmotic system. However, in contrast to Equation (10.2), the same pressure $p^{(1)}$ appears in both volume work terms. The failure to recognize this difference has been responsible for much confusion in the treatment of surface phenomena.

In accordance with the standard procedure, the next step involves the construction of a function of state whose differential along some path in state space *imitates* $D\varphi$. One such function is

$$J = U + p^{(1)}V^{(1)} + p^{(1)}V^{(2)} - TS. \tag{11.13}$$

Consider a path of constant T and $p^{(1)}$. Along this path

$$[D\varphi]_{T,p^{(1)}} = \delta U + p^{(1)}\,dV + p^{(1)}\,dV^{(2)} - T\,\delta S, \tag{11.14}$$

while

$$[\delta J]_{T,p^{(1)}} = \delta U + p^{(1)}\,\delta V + p^{(1)}\,\delta V^{(2)} - T\,\delta S. \tag{11.15}$$

Comparison of Equations (11.14) and (11.15) shows that δJ imitates $D\varphi$ along this path. Thus, J may be used as a thermodynamic potential and the extremal condition,

$$[\delta J]_{T,p^{(1)}} = 0, \tag{11.16}$$

may be employed as a "feature" of equilibrium.

Introduce the relations

$$\begin{aligned} S &= S^{(1)} + S^{(2)}, \\ U &= U^{(1)} + U^{(2)}, \end{aligned} \tag{11.17}$$

in which both the entropy and internal energy of the surface are included in the respective quantities $S^{(2)}$ and $U^{(2)}$ for the drop (in keeping with the assignment of the surface to the drop). Then,

$$J = (U^{(1)} + p^{(1)}V^{(1)} - TS^{(1)}) \\ + (U^{(2)} + p^{(1)}V^{(2)} - TS^{(2)}) = G^{(1)} + R^{(2)}. \tag{11.18}$$

Notice that the symbol $G^{(2)}$ is not employed for the second quantity in brackets but that $R^{(2)}$ is used instead. This is because, although the first quantity in brackets *is* a Gibbs free energy and may be denoted by $G^{(1)}$, the second quantity *is not*, for it contains the pressure $p^{(1)}$ of phase 1 rather than $p^{(2)}$, the pressure of the phase to which it refers. Furthermore, it should also be noted that the thermodynamic potential J is not a Gibbs function even though one of its component terms, $G^{(1)}$, is.

4. An Alternative Form for $R^{(2)}$

It will now be demonstrated that

$$R^{(2)}(T,p^{(2)}) = [G^{(2)}(T,p^{(1)})]_{\text{bulk}} + \sigma\Omega, \tag{11.19}$$

where $R^{(2)}(T,p^{(2)})$ is simply the $R^{(2)}$ defined in Equation (11.18), that is,

$$R^{(2)}(T,p^{(2)}) = U^{(2)}(T,p^{(2)}) + p^{(1)}V^{(2)}(T,p^{(2)}) - TS^{(2)}(T,p^{(2)}), \tag{11.20}$$

and $[G^{(2)}(T,p^{(1)})]_{\text{bulk}}$ is simply the Gibbs free energy, which all of the material in the drop would possess were it part of a bulk phase at the temperature T and pressure $p^{(1)}$. In Equation (11.20) the arguments $(T,p^{(2)})$ are shown for each extensive quantity to remind the reader that even though $p^{(1)}$ appears in $R^{(2)}$, $R^{(2)}$ itself is a function associated with a system existing at the local pressure $p^{(2)}$. Furthermore, it should be recalled that the surface is part of that system so that both $U^{(2)}(T,p^{(2)})$ and $S^{(2)}(T,p^{(2)})$ contain surface contributions.

The form (11.19) is actually the more common one for $R^{(2)}$. However, its connection with fundamentals is more obscure than Equation (11.18), and it is often employed in an incautious and erroneous manner. One of the confusions arising in connection with Equation (11.19) lies in the failure to realize that $G^{(2)}$ is the bulk-free energy of the material in the drop at the pressure $p^{(1)}$ *outside* rather than at $p^{(2)}$ *inside* the drop. Other confusions have been generated by the usual insistence that $R^{(2)}$ be regarded as an augmented Gibbs free energy when it is clear that there are fundamental differences between it and the Gibbs function in the usual sense. We shall discuss an example below. Now back to our proof.

For $[G^{(2)}(T,p^{(1)})]_{\text{bulk}}$ we may write, according to Equation (8.25),

$$[G^{(2)}(T,p^{(1)})]_{\text{bulk}} = [U^{(2)}(T,p^{(1)})]_{\text{bulk}} + p^{(1)}[V^{(2)}(T,p^{(1)})]_{\text{bulk}} - T[S^{(2)}(T,p^{(1)})]_{\text{bulk}}. \quad (11.21)$$

By combining Equations (11.20) and (11.21) we get†

$$R^{(2)}(T,p^{(2)}) - [G^{(2)}(T,p^{(1)})]_{\text{bulk}} = U^{(2)}(T,p^{(2)}) - [U^{(2)}(T,p^{(1)})]_{\text{bulk}} - T\{S^{(2)}(T,p^{(2)}) - [S^{(2)}(T,p^{(1)})]_{\text{bulk}}\} = \Delta U - T\,\Delta S = -\Delta w. \quad (11.22)$$

Several manipulations have been included in Equation (11.22). The expression on the right of the first equals sign is obtained by subtracting Equation (11.21) from Equation (11.20), making use of the fact that

$$V^{(2)}(T,p^{(2)}) = [V^{(2)}(T,p^{(1)})]_{\text{bulk}}, \quad (11.23)$$

which comes from the assumption that the liquid is incompressible. ΔU and ΔS are obviously the changes in internal energy and entropy corresponding to the change of state in which a drop is formed by extracting material from bulk liquid at pressure $p^{(1)}$. The expression $\Delta U - T\,\Delta S$ following the second equals sign in Equation (11.22) is obviously (see Equations (3.18) and (4.38)) the work $-\Delta w$ performed *on* the system when this process is conducted reversibly.

In order to prove Equation (11.19) then it is only necessary to identify $-\Delta w$ with $\sigma\Omega$. This may be accomplished with the help of the apparatus exhibited in Figure 11.3. In the figure, the bulk liquid is enclosed in the barrel

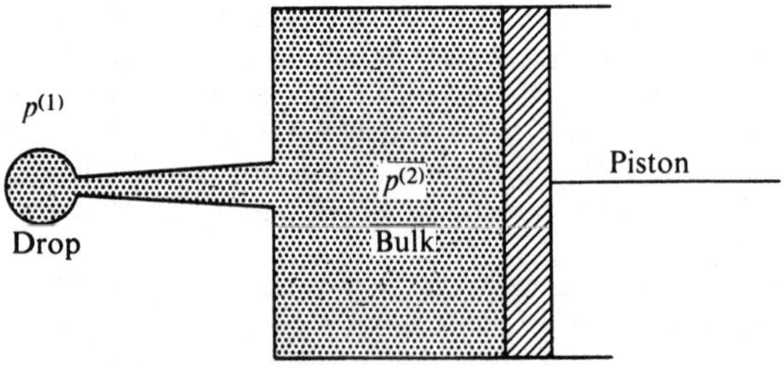

FIGURE 11.3. *Apparatus for fashioning a drop from bulk liquid.*

† Notice that $R^{(2)}(T,p^{(1)})$ is not the same as $[G^{(2)}(T,p^{(1)})]_{\text{bulk}}$ because $R^{(2)}$ contains surface contributions.

of an instrument resembling a syringe whose hollow needle tapers to an infinitely fine point. Through application of pressure by means of the piston, a spherical drop can be formed at the end of the needle. The process is conducted quasistatically so that the pressure $p^{(1)}$ of the vapor which surrounds the drop and syringe is always the equilibrium vapor pressure of the drop. In this manner, material from the bulk phase may be fashioned into a drop.

Now at first the piston only subjects the bulk liquid to the pressure $p^{(1)}$ so that we may start with the desired initial state. In order to form the drop, it is necessary for the piston to apply a pressure slightly in excess of $p^{(2)}$, where $p^{(2)}$ is given by Equation (11.11). Since the liquid is incompressible and no motion occurs until $p^{(2)}$ is reached, no work is performed while the pressure is being increased.

Actually, if $p^{(2)}$ were prescribed by Equation (11.11) down to values of r approaching zero, $p^{(2)}$ would initially have to be infinite since r is initially zero. Physically, Equation (11.11) breaks down before r reaches zero, and so in practice this catastrophe would be avoided. However, even if (11.11) is applied at small values of r, no real divergence occurs because the integral with which we deal below, Equation (11.27), converges.

Once the piston begins to move, the reversible work $-\Delta w$ may be computed as follows. The work performed by the piston on the liquid is

$$p^{(2)}\,dV^{(2)}, \tag{11.24}$$

where $dV^{(2)}$ is the increment in the volume of the drop. But the drop performs work on *its* environment because it displaces vapor at pressure $p^{(1)}$. This work amounts to

$$p^{(1)}\,dV^{(2)}. \tag{11.25}$$

Thus, the net work performed on the liquid is

$$(p^{(2)} - p^{(1)})\,dV^{(2)}, \tag{11.26}$$

and for the complete process,

$$-\Delta w = \int_0^V (p^{(2)} - p^{(1)})\,dV^{(2)}. \tag{11.27}$$

The use of Equation (11.10) for $dV^{(2)}$ and Equation (11.11) for the integrand gives

$$-\Delta w = \int_0^r \left(\frac{2\sigma}{r}\right) 4\pi r^2\,dr = \sigma 4\pi r^2 = \sigma\Omega, \tag{11.28}$$

where Equation (11.09) has been used.

5. The Chemical Potential

In place of Equation (11.18) we may now write

$$J = G^{(1)}(T,p^{(1)}) + [G^{(2)}(T,p^{(1)})]_{\text{bulk}} + \sigma\Omega. \tag{11.29}$$

This may be substituted into Equation (11.16) to investigate the condition of internal equilibrium in the system consisting of the drop and its vapor. $G^{(1)}$ and $[G^{(2)}]_{\text{bulk}}$ depend upon T, $p^{(1)}$ and $n^{(1)}$, and T, $p^{(1)}$ and $n^{(2)}$, respectively, where $n^{(1)}$ and $n^{(2)}$ are the respective mole numbers. Since the system is closed,

$$\delta n^{(2)} = -\delta n^{(1)}. \tag{11.30}$$

Use of Equation (11.16) now gives

$$\begin{aligned}[\delta J]_{T,p^{(1)}} &= \left[\frac{\partial G^{(1)}}{\partial n^{(1)}}\right]_{T,p^{(1)}} \delta n^{(1)} + \left[\frac{\partial}{\partial n^{(2)}}\{[G^{(2)}]_{\text{bulk}} + \sigma\Omega\}\right]_{T,p^{(1)}} \delta n^{(2)} \\ &= \left\{\mu^{(1)} - \left[\frac{\partial R^{(2)}}{\partial n^{(2)}}\right]_{T,p^{(1)}}\right\} \delta n^{(1)} = 0,\end{aligned} \tag{11.31}$$

where Equation (11.30) and the Definitions (8.57) and (11.19) have been used. Thus, for the condition of internal equilibrium we have

$$\mu^{(1)} = \left[\frac{\partial R^{(2)}}{\partial n^{(2)}}\right]_{T,p^{(1)}}. \tag{11.32}$$

This equation states that for equilibrium, with respect to the transfer of matter between the vapor and the drop, the chemical potential of the vapor should equal the derivative on the right. This derivative which is associated entirely with the drop, therefore plays the role of a chemical potential for the drop. With its help, some of the confusion, generated when derivations are based upon analogy rather than upon Equation (8.33), can be illustrated. We discuss one example.

Equation (11.32) applies to a one-component system. In particular, the drop consists of a single component. For an ordinary one-component phase, Equation (3.31) requires

$$\mu = \left[\frac{\partial G}{\partial n}\right]_{T,p} = \frac{G}{n}. \tag{11.33}$$

If in analogy with Equation (11.33) one considers $R^{(2)}$ to behave like an ordinary free energy, he might be tempted to define the chemical potential of the drop as

$$\mu^{(2)} = \frac{R^{(2)}}{n^{(2)}}. \tag{11.34}$$

In accordance with Equation (11.33), this would be true if $\frac{R^{(2)}}{\hbar^{(2)}}$ were equal to $\left[\frac{\partial R^{(2)}}{\partial n^{(2)}}\right]_{T,p^{(1)}}$. This type of relation holds for G in Equation (11.33), but unfortunately, not for $R^{(2)}$; that is,

$$\frac{R^{(2)}}{n^{(2)}} \neq \left[\frac{\partial R^{(2)}}{\partial n^{(2)}}\right]_{T,p^{(1)}}, \tag{11.35}$$

in the system to which Equation (11.32) applies. Whereas G in Equation (11.33) is a homogeneous function in the first degree in n, $R^{(2)}$ is not homogeneous in $n^{(2)}$. Only the term $[G^{(2)}]_{\text{bulk}}$ is homogeneous, whereas $\sigma\Omega$ is proportional to $[n^{(2)}]^{2/3}$. Thus, the *derivative* of $R^{(2)}$ with respect to $n^{(2)}$ *is* a chemical potential, but the *ratio* of $R^{(2)}$ to $n^{(2)}$ *is not*.

A number of distinguished investigators have published papers in which they assume that $\mu^{(2)}$ defined by Equation (11.34) is the proper chemical potential. In committing this error, reasoning is usually performed by analogy rather than through a return to the fundamentals embodied in Equation (8.33). This example illustrates how important it is to return to Equation (8.33) each time a new *kind* of thermodynamic system is treated.

6. Dependence of Vapor Pressure on Drop Size

As in previous chapters, the further implications of a relation like Equation (11.32), involving the equality of chemical potentials, can only be approached through a knowledge of how both $\mu^{(1)}$ and $\left[\dfrac{\partial R^{(2)}}{\partial n^{(2)}}\right]_{T,p^{(1)}}$ depend upon the variables of state. In this section, we shall call upon such knowledge to determine the manner in which the vapor pressure of the drop depends upon radius r. We shall assume, as we have up to now, that the liquid comprising the drop is incompressible. Furthermore, it will also be assumed that the surface tension σ is independent of r. These conditions cannot be arrived at by thermodynamic reasoning and are to be regarded simply as constituting empirical data obtained in connection with this particular system. The specialization, however, has the merit of permitting us to demonstrate the method with a minimum of manipulative complexity.

If Equation (11.19) is substituted into Equation (11.32) the result is

$$\mu^{(1)}(T,p^{(1)}) = \mu^{(2)}(T,p^{(1)}) + \sigma\left[\frac{\partial\Omega}{\partial n^{(2)}}\right]_{T,p^{(1)}}, \tag{11.36}$$

where Equation (8.57) and the fact that σ is independent of drop size has been utilized. In general, the equilibrium vapor pressure is $p^{(1)}$, but for the special case $r \to \infty$, that is, for bulk liquid we may use the symbol $p_\infty^{(1)}$. Now for bulk liquid in equilibrium with its vapor, Equation (8.49) must hold so that

$$\mu^{(1)}(T,p_\infty^{(1)}) = \mu^{(2)}(T,p_\infty^{(1)}). \tag{11.37}$$

Furthermore,

$$\begin{aligned}\mu^{(1)}(T,p^{(1)}) - \mu^{(1)}(T,p_\infty^{(1)}) &= \int_{p_\infty^{(1)}}^{p^{(1)}} \left[\frac{\partial\mu^{(1)}}{\partial p^{(1)}}\right]_{T,n^{(1)}} dp^{(1)} \\ &= \int_{p_\infty^{(1)}}^{p^{(1)}} \bar{v}^{(1)}\, dp^{(1)},\end{aligned} \tag{11.38}$$

where Equation (5.66) has been used. Similarly,

$$\mu^{(2)}(T,p^{(1)}) - \mu^{(2)}(T,p_\infty^{(1)}) = \int_{p_\infty^{(1)}}^{p^{(1)}} \bar{V}^{(2)}\, dp^{(1)}. \tag{11.39}$$

Substitution of Equations (11.38) and (11.39) into Equation (11.36) gives

$$\mu^{(1)}(T,p_\infty^{(1)}) + \int_{p_\infty^{(1)}}^{p^{(1)}} \bar{V}^{(1)}\, dp^{(1)} = \mu^{(2)}(T,p_\infty^{(1)}) + \int_{p_\infty^{(1)}}^{p^{(1)}} \bar{V}^{(2)}\, dp^{(1)} + \sigma\left[\frac{\partial\Omega}{\partial n^{(2)}}\right]_{T,p^{(1)}}, \tag{11.40}$$

and application of Equation (11.37) simplifies this to

$$\int_{p_\infty^{(1)}}^{p^{(1)}} \bar{V}^{(1)}\, dp^{(1)} = \int_{p_\infty^{(1)}}^{p^{(1)}} \bar{V}^{(2)}\, dp^{(1)} + \sigma\left[\frac{\partial\Omega}{\partial n^{(2)}}\right]_{T,p^{(1)}}. \tag{11.41}$$

We make the further assumption that the vapor phase behaves like an ideal gas (a good approximation at the densities common to most vapors). Then Equations (3.31) and (5.1) may be used to arrive at the following result:

$$\bar{V}^{(1)} = \frac{V^{(1)}}{n^{(1)}} = \frac{RT}{p^{(1)}}. \tag{11.42}$$

For the liquid phase, considered incompressible, we have

$$\bar{V}^{(2)} = \frac{V^{(2)}}{n^{(2)}} = V^{(2)0}, \tag{11.43}$$

where $V^{(2)0}$ is a constant and represents the molar volume of the liquid. Substitution of Equations (11.42) and (11.43) into Equation (11.41) yields

$$RT \ln \frac{p^{(1)}}{p_\infty^{(1)}} = V^{(2)0}(p^{(1)} - p_\infty^{(1)}) + \sigma\left[\frac{\partial\Omega}{\partial n^{(2)}}\right]_{T,p^{(1)}}. \tag{11.44}$$

Now according to Equation (11.9),

$$n^{(2)}V^{(2)0} = V^{(2)} = \tfrac{4}{3}\pi r^3, \tag{11.45}$$

from which we obtain

$$r = \left\{\frac{3n^{(2)}V^{(2)0}}{4\pi}\right\}^{1/3}, \tag{11.46}$$

and when this is substituted into the second of equations (11.9), there is obtained

$$\Omega = 4\pi r^2 = 4\pi\left\{\frac{3n^{(2)}V^{(2)0}}{4\pi}\right\}^{2/3}. \tag{11.47}$$

From this,

$$\left[\frac{\partial\Omega}{\partial n^{(2)}}\right]_{T,p^{(1)}} = 2V^{(2)0}\left\{\frac{3n^{(2)}V^{(2)0}}{4\pi}\right\}^{-1/3} = \frac{2V^{(2)0}}{r}. \tag{11.48}$$

Equation (11.48) may be substituted into Equation (11.44) to yield

$$RT\ln\frac{p^{(1)}}{p_{\infty}^{(1)}} = v^{(2)0}[p^{(1)} - p_{\infty}^{(1)}] + \frac{2\sigma v^{(2)0}}{r}. \tag{11.49}$$

Usually the first term on the right is so small that it may be neglected, and the more familiar result

$$RT\ln\frac{p^{(1)}}{p_{\infty}^{(1)}} = \frac{2\sigma v^{(2)0}}{r} \tag{11.50}$$

is obtained. This equation shows how the vapor pressure $p^{(1)}$ of the drop depends upon its radius r.

7. Dependence of Solubility on Drop Size

If the drop is surrounded, not by its vapor but by a bulk, binary solution in which the drop substance is a solute, and such that the solvent is itself insoluble in the drop, the drop continues to consist of a single component, but is now in equilibrium with a solution. It becomes a valid question to inquire into the solubility of the drop in the surrounding solution, and particularly into how this solubility depends upon r. If the solution is ideal or dilute so that Equation (9.1) or Equation (9.47) applies, this dependence may be derived by an argument very similar to the one presented in the preceding section. The resulting formula is

$$RT\ln\frac{x_2^{(1)}}{x_{2\infty}^{(1)}} = \frac{2\sigma v^{(2)0}}{r}, \tag{11.51}$$

where σ is now the interfacial tension between the drop and the solution. Equation (11.51) is almost identical with Equation (11.50) except for the fact that the ratio $x_2^{(1)}/x_{2\infty}^{(1)}$ appears in place of $p^{(1)}/p_{\infty}^{(1)}$. The quantity $x_2^{(1)}$ is the mole fraction of solute (drop substance) in the solution at equilibrium, that is, the solubility, and $x_{2\infty}^{(1)}$ is clearly the solubility when $r \to \infty$.

8. Surface Effects and Structure of the Surface Layer

In the second stage of refinement, mentioned in Section 1 of this chapter, one can no longer look upon the interface between two phases as a mathematical plane of zero thickness at which the properties of one phase pass discontinuously into the properties of another. Instead, it is necessary to take account of the manner in which the sharp but continuous transition in properties actually occurs. In this section, we shall confine our remarks to systems with planar interfacial layers. For illustrative purposes one may

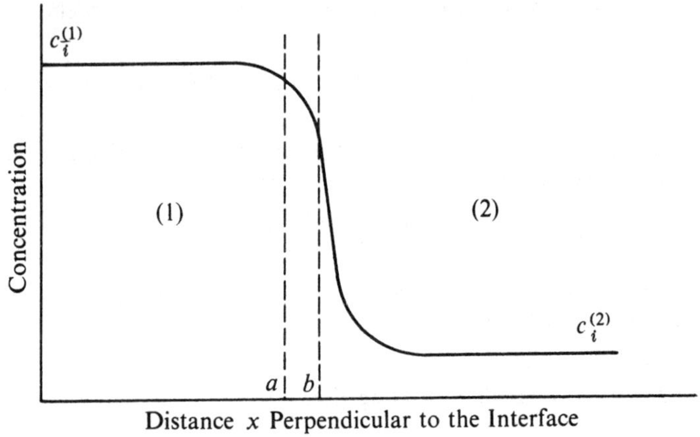

FIGURE 11.4. *Change of concentration at an interface between two phases. The dashed lines represent two different choices of dividing surface.*

consider Figure 11.4 which might apply to a two-phase system (phases 1 and 2) with a plane interface. In the Figure we have plotted the concentration of the ith component as a function of the distance x perpendicular to the interface. Although a plane of true discontinuity does not exist, one can nevertheless choose arbitrarily a mathematical plane of reference located, say at $x = a$, in the Figure. Such a plane will be referred to as a *Gibbs dividing surface.* Far from the dividing surface (on either side) the concentration becomes uniform as the homogeneous phases are penetrated. Upon employing the conventional notation, these uniform concentrations may be denoted as $c_i^{(1)}$ and $c_i^{(2)}$, respectively.

The dividing surface partitions the total volume V of the system into two distinct parts, $V^{(1)}$ and $V^{(2)}$,

$$V = V^{(1)} + V^{(2)}. \tag{11.52}$$

If the phases were homogeneous throughout $V^{(1)}$ and $V^{(2)}$, that is, uniform right up to the dividing surface, the total number of moles of the ith species in the system would be

$$n_i^* = c_i^{(1)}V^{(1)} + c_i^{(2)}V^{(2)} = n_i^{(1)} + n_i^{(2)}, \tag{11.53}$$

where $n_i^{(1)}$ and $n_i^{(2)}$ are the mole numbers of the ith species which would be present. However, because the phases are not perfectly homogeneous, the true number of moles is some other number n_i. A consistent treatment may be advanced by focusing attention upon the deviations from this ideal situation. Thus, the quantity,

$$n_i^{(s)} = n_i - n_i^* = n_i - n_i^{(1)} - n_i^{(2)}, \tag{11.54}$$

may be defined as the number of moles of the ith species *adsorbed* at the interface. It is clear from the definitions that $n_i^{(s)}$ is simply the excess of the actual number of moles of ith species in the system over the number which would be present if the system were perfectly uniform, on each side, up to the dividing surface. The assemblage of *excess* extensive variables of this kind can be used to define a new *excess* or *surface* phase. In order for the treatment to remain valid, it is necessary to perform all the bookkeeping accurately so that the inventories of energy and matter in the system always add up to the observed values.

A surface concentration $\Gamma_i^{(s)}$ may be defined by dividing $n_i^{(s)}$ by Ω, the area of the interface. Thus,

$$\Gamma_i^{(s)} = \frac{n_i^{(s)}}{\Omega} . \tag{11.55}$$

$\Gamma_i^{(s)}$ may be named the *surface concentration* of the ith species or, as it is sometimes called, the *superficial density* of the ith species. It is evident from geometric considerations that both $\Gamma_i^{(s)}$ and $n_i^{(s)}$ will depend upon the exact choice of dividing surface. Thus, if the plane at $x = b$ in Figure 11.4 had been chosen instead of the one at $x = a$, the numerical values of $\Gamma_i^{(s)}$ and $n_i^{(s)}$ would be different. Therefore, as part of the thermodynamic treatment the choice of dividing surface must be specified.

If this specification required a detailed knowledge of the distribution of matter in the interfacial region, say over a distance of 20 angstroms, there might be some question concerning whether or not the treatment was being advanced in a manner sufficiently macroscopic to qualify as a thermodynamic one. Fortunately, there exist valuable applications of the present analysis which do not require so detailed a knowledge of the interfacial region, and in connection with which $\Gamma_i^{(s)}$ may be determined by macroscopic means.

It is convenient to choose the dividing surface so that the surface concentration of the component designated as the solvent is zero. It is customary to denote the solvent by the subscript 1. Therefore, this choice is specified by the condition

$$\Gamma_1^{(s)} = 0. \tag{11.56}$$

We shall adopt this convention.

9. Excess Quantities in General

Other excess quantities besides $n_i^{(s)}$ may be defined immediately. By definition, the mole numbers of a unit volume of any homogeneous phase are the concentrations c_i. Thus, the mole numbers in unit volume of phase 1 (away from the interface) are $c_i^{(1)}$. In accordance with Equation (3.30) then, the value of any extensive quantity $E^{(1)}$ in unit volume (the *density* $e^{(1)}$ of that quantity) of phase 1 is given by

$$e^{(1)} = \sum_i c_i^{(1)} \bar{E}_i^{(1)}, \tag{11.57}$$

where $\bar{E}_i^{(1)}$ is a partial molar quantity.

Thus, the density $u^{(1)}$ of the internal energy is

$$u^{(1)} = \sum_i c_i^{(1)} \bar{U}_i^{(1)}. \tag{11.58}$$

If $U^{(1)}$ is defined (in analogy to $n_i^{(1)}$) as the internal energy which phase 1 would possess if it were uniform right up to the dividing surface, then

$$U^{(1)} = V^{(1)} u^{(1)}, \tag{11.59}$$

with a similar relation for $U^{(2)}$.

If it is assumed that the internal energy density is uniform up to the dividing surface, the excess internal energy $U^{(s)}$ is given (in analogy to Equation (11.54)) by

$$\begin{aligned} U^{(s)} &= U - U^{(1)} - U^{(2)} = U - V^{(1)}u^{(1)} - V^{(2)}u^{(2)} \\ &= U - \sum_i (c_i^{(1)} V^{(1)})\bar{U}_i^{(1)} - \sum_i (c_1^{(2)} V^{(2)})\bar{U}_i^{(2)} \\ &= U - \sum_i n_i^{(1)}\bar{U}_i^{(1)} - \sum_i n_i^{(2)}\bar{U}_i^{(2)}, \end{aligned} \tag{11.60}$$

where U is the actual internal energy of the system and where Equations (11.53), (11.58), and (11.59) have been used. Other excess quantities, for example $S^{(s)}$, $G^{(s)}$, et cetera, are defined in precisely the same way.

10. Surface Constraints and Surface Tension

Next consider the differential of the internal energy for the entire system. In accordance with Equations (8.1) and (4.38), we may write

$$dU = T\,dS - Dw_{0_{\text{rev}}}. \tag{11.61}$$

For the reversible work performed during this change, we have written $Dw_{0_{\text{rev}}}$ because for the moment we wish to consider changes of state brought about by varying the set of constraints χ_0.

Later, when a suitable thermodynamic potential has been defined, χ_0 will be fixed and a variational constraint χ_1 will be introduced to drive the system. At first, however, we will permit χ_0 to vary so that the dependence

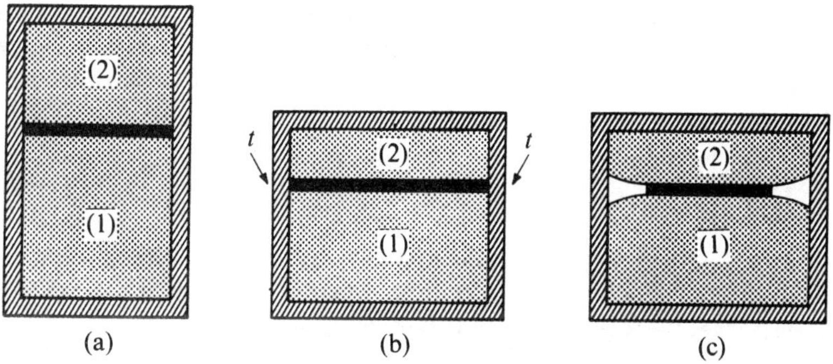

FIGURE 11.5. *Different methods of constraining a surface with a single interface.*

of U on the variables of state which correspond to these constraints may be investigated, and the differential coefficients of Equation (11.61) may be properly identified.

As usual, the constraints determine the *kind* of work the system may do, and the next step is the exploration of this question. In Figure 11.1, the system is arranged so that it may perform only volume work on its surroundings, even though it does contain an interface. This interface is closed upon itself and does not contact the environment. In the present section, our interest is focused upon a plane interface, and we are therefore faced with a configuration something like that illustrated in Figure 11.5(*a*). A two-phase system (for convenience assume that it is liquid-liquid, for example benzene-water) is shown with the plane interface exhibited as a heavy horizontal line. Phases 1 and 2 are marked. The entire system is in a container (cross hatched). If the shape of the system is changed without necessarily changing its volume, the process might terminate in the configuration, Figure 11.5(*b*), in which the area of the interface has been increased. In this case, the environment would have performed *surface* work upon the system. It is clear that, in contrast to the case of Figure 11.1, the environment now makes direct contact with the interface at the points t in Figure 11.5(*b*) where it meets the wall. Thus, among the constraints, χ_0 is the condition that the interface always meet the walls of the containing vessel.

Alternatively, we may have a situation, such as the one illustrated in Figure 11.5(*c*), where the interface breaks loose from the wall so that the surroundings cannot perform the same surface work as with Figure 11.5(*b*). In fact, new kinds of surfaces are produced, for example vapor-liquid interfaces. Furthermore, they will be curved so that the restriction to planarity will no longer hold. Since we have decided to treat plane interfaces, it is apparent that attention must be confined to processes such as the one leading from

Figure 11.5(*a*) to 11.5(*b*). Therefore, attachment to walls at points t must be included among the constraints χ_0; and the system, in contrast to Figure 11.1, must be able to exchange surface work directly with the environment. As a result, one may write Equation (11.61) more explicitly in the form

$$dU = T\,dS - p\,dV + \left[\frac{\partial U}{\partial \Omega}\right]_{S,V} d\Omega, \tag{11.62}$$

where the work term $Dw_{0_{\text{rev}}}$ is now seen to contain, besides the usual volume work $p\,dV$, a term proportional to the increment $d\Omega$ in surface area. This is the surface work term. It is customary to define

$$\sigma = \left[\frac{\partial U}{\partial \Omega}\right]_{S,V} \tag{11.63}$$

as the surface tension. Clearly, from the position of the term $\sigma\,d\Omega$ in Equation (11.62), it represents the surface work performed *on* the system when the surface area is increased reversibly by $d\Omega$, and therefore σ is the reversible work which must be expended during a unit increase in surface area Ω. This definition of σ is equivalent to the one given in Section 2 of this chapter. Since the composite system is closed, the mole numbers of all species remain fixed during the change to which Equation (11.62) corresponds, although the distribution of each species between the bulk phases and the interfacial layer may change. Strictly speaking then, the subscripts in Equation (11.63) should contain these mole numbers. For simplicity we have omitted them.

11. The Thermodynamic Potential

In Equation (11.62) it has been assumed that the temperature is uniformly T throughout the system (maintained so by a thermostat), and that in view of the absence of curved surfaces the pressure is uniformly p. For future reference we write $Dw_{0_{\text{rev}}}$

$$Dw_{0_{\text{rev}}} = p\,dV - \sigma\,d\Omega. \tag{11.64}$$

Later, when the thermodynamic potential is derived through use of Equation (8.33), we shall have occasion to employ this equation. It must be emphasized again that the specific form assumed by $Dw_{0_{\text{rev}}}$ is sensitive to the constraints χ_0. If Figure 11.5(*c*) applied, Equation (11.64) would not be valid. Neither is it valid for the configuration of Figure 11.1.

In view of the definitions of $U^{(1)}$, $U^{(2)}$, and $U^{(s)}$, one may write Equation (11.62) in the form

$$dU^{(1)} + dU^{(2)} + dU^{(s)} = T\,dS - p\,dV + \sigma\,d\Omega. \tag{11.65}$$

Now even though the *total* system is closed, the individual subsystems—phases 1 and 2 and the surface excess region—may be *open* so that $n_i^{(1)}$, $n_i^{(2)}$, and $n_i^{(s)}$ may vary. In this case, the differential terms on the left of Equation (11.65) may contain differentials of these composition variables. For the time being, however, assume that the composition of each subsystem is fixed. Then if we write

$$dU^{(1)} = T\,dS^{(1)} - p\,dV^{(1)}, \tag{11.66}$$

$$dU^{(2)} = T\,dS^{(2)} - p\,dV^{(2)}, \tag{11.67}$$

$$dU^{(s)} = T\,dS^{(s)} + \sigma\,d\Omega, \tag{11.68}$$

it follows from Equation (11.52) and the Definitions (11.59) of $U^{(1)}$ and $U^{(2)}$ (and also, by the same procedure, of $S^{(1)}$ and $S^{(2)}$) as well as the definitions of $U^{(s)}$ and $S^{(s)}$ by Equation (11.60), that the sum of these equations reproduces Equations (11.65). Therefore, Equation (11.68) is at least a consistent form. The surface excess region to which it applies is a construct resembling the last entry of an accounting operation. The books are balanced by arbitrarily forcing liabilities to equal assets. The dependence upon composition may be introduced in the usual manner. Thus,

$$dU^{(1)} = T\,dS^{(1)} - p\,dV^{(1)} + \sum_i \mu_i^{(1)}\,dn_i^{(1)}, \tag{11.69}$$

$$dU^{(2)} = T\,dS^{(2)} - p\,dV^{(2)} + \sum_i \mu_i^{(2)}\,dn_i^{(2)}, \tag{11.70}$$

$$dU^{(s)} = T\,dS^{(s)} + \sigma\,d\Omega + \sum_i \mu_i^{(s)}\,dn_i^{(s)}. \tag{11.71}$$

Here $\mu_i^{(1)}$ and $\mu_i^{(2)}$, defined as they are for homogeneous phases, are the usual chemical potentials as defined by Equation (8.57), while $\mu_i^{(s)}$ is given by

$$\mu_i^{(s)} = \left[\frac{\partial U^{(s)}}{\partial n_i^{(s)}}\right]_{S,\Omega} \tag{11.72}$$

in view of its position in Equation (11.71). In effect, Equation (11.71) implies that the independent variables of the construct which we have called the surface excess region are S, Ω, and $n_i^{(s)}$. As in the case of any thermodynamic system, assumptions must be made concerning what constitutes a set of state variables, and the choice must be justified by experiment.

Upon returning to Equation (8.33) and introducing Equation (11.64), we obtain

$$D\varphi = \delta U - T\,\delta S + p\,\delta V - \sigma\,\delta\Omega = 0. \tag{11.73}$$

Once again a thermodynamic potential may be constructed by synthesizing a function of state whose differential *imitates* $D\varphi$ over a selected path in state space. Consider the following function:

$$L = U + pV - \sigma\Omega - TS. \tag{11.74}$$

Along a path of constant T, p, and σ,

$$(\delta L)_{T,p,\sigma} = \delta U - T\,\delta S + p\,\delta V - \sigma\,\delta\Omega, \tag{11.75}$$

while

$$(D\varphi)_{T,p,\sigma} = \delta U - T\,\delta S + p\,\delta V - \sigma\,\delta\Omega. \tag{11.76}$$

Hence, δL *imitates* $D\varphi$ over this path. Thus, we may employ L as a thermodynamic potential. The extremal condition becomes

$$(\delta L)_{T,p,\sigma} = 0. \tag{11.77}$$

By substituting Equation (11.52), together with Equation (11.60) and its equivalent for S, into Equation (11.74) we get,

$$L = [U^{(1)} + pV^{(1)} - TS^{(1)}] + [U^{(2)} + pV^{(2)} - TS^{(2)}] + [U^{(s)} - \sigma\Omega - TS^{(s)}] = G^{(1)} + G^{(2)} + Z^{(s)}, \tag{11.78}$$

where $G^{(1)}$ and $G^{(2)}$ are Gibbs free energies (see Equation (4.71)) and

$$Z^{(s)} = U^{(s)} - \sigma\Omega - TS^{(s)}. \tag{11.79}$$

Neither L nor $Z^{(s)}$ is a Gibbs function. Both $G^{(1)}$ and $G^{(2)}$ like $U^{(1)}$, $V^{(1)}$, $S^{(1)}$, and $U^{(2)}$, $V^{(2)}$, $S^{(2)}$ correspond to homogeneous phases which are imagined to extend uniformly right up to the dividing surface.

12. The Chemical Potential

The extremal condition now assumes the form

$$(\delta L)_{T,p,\sigma} = (\delta G^{(1)})_{T,p} + (\delta G^{(2)})_{T,p} + (\delta Z^{(s)})_{T,\sigma} = 0. \tag{11.80}$$

The usual Legendre transformations are performed on $dG^{(1)}$ and $dG^{(2)}$. Thus,

$$dG^{(1)} = dU^{(1)} + p\,dV^{(1)} + V^{(1)}\,dp - S^{(1)}\,dT - T\,dS^{(1)}. \tag{11.81}$$

Substitution from Equation (11.69) yields,

$$dG^{(1)} = V^{(1)}\,dp - S^{(1)}\,dT + \sum_i \mu_i^{(1)}\,dn_i^{(1)}, \tag{11.82}$$

and a similar relation may be derived for $dG^{(2)}$. In the case of $dZ^{(s)}$ we write

$$dZ^{(s)} = dU^{(s)} - \sigma\,d\Omega - \Omega\,d\sigma - T\,dS^{(s)} - S^{(s)}\,dT. \tag{11.83}$$

Substitution of Equation (11.71) into this equation gives

$$dZ^{(s)} = -\Omega\,d\sigma - S^{(s)}\,dT + \sum_i \mu_i^{(s)}\,dn_i^{(s)}. \tag{11.84}$$

If now Equation (11.82) and the similar relation for $dG^{(2)}$ are introduced, together with Equation (11.84) into Equation (11.80), the result is

$$\sum_i \mu_i^{(1)}\,\delta n_i^{(1)} + \sum_i \mu_i^{(2)}\,\delta n_i^{(2)} + \sum_i \mu_i^{(s)}\,\delta n_i^{(s)} = 0. \tag{11.85}$$

Since the composite system is closed, n_i in Equation (11.54) is constant. Thus, from Equation (11.54),

$$\delta n_i^{(1)} + \delta n_i^{(2)} + \delta n_i^{(s)} = 0, \qquad i = 1, 2, \ldots, C. \tag{11.86}$$

Multiplication of Equation (11.86) by an undetermined multiplier λ_i as in Chapter 8, Section 9, and addition of the result to Equation (11.85), results in

$$\sum_i (\mu_i^{(1)} + \lambda_i)\,\delta n_i^{(1)} + \sum_i (\mu_i^{(2)} + \lambda_i)\,\delta n_i^{(2)} + \sum_i (\mu_i^{(s)} + \lambda_i)\,\delta n_i^{(s)} = 0. \tag{11.87}$$

The coefficients must vanish identically (see Chapter 2, Section 6) so that

$$\mu_i^{(1)} = \mu_i^{(2)} = \mu_i^{(s)} = -\lambda_i. \tag{11.88}$$

Thus, the condition of internal equilibrium requires that $\mu_i^{(s)}$ equal the chemical potentials in phases 1 and 2. Thus, $\mu_i^{(s)}$, defined by Equation (11.72), plays the role of a chemical potential.

13. The Gibbs-Duhem Relation for the Surface

Now both σ and T in Equation (11.80) are intensive quantities, whereas $Z^{(s)}$ is clearly extensive. Thus, $Z^{(s)}$ is a homogeneous function of the first degree in the surface excess mole numbers $n_i^{(s)}$, provided that σ and T are held constant. Euler's theorem (Equation (2.43) then requires

$$Z^{(s)} = \sum_i n_i^{(s)} \left[\frac{\partial Z^{(s)}}{\partial n_i^{(s)}}\right]_{T,\sigma,n^{(s)}}. \tag{11.89}$$

An examination of Equation (11.84) shows a definition for $\mu_i^{(s)}$, alternative but equivalent to the one presented in Equation (11.72). This is

$$\mu_i^{(s)} = \left[\frac{\partial Z^{(s)}}{\partial n_i^{(s)}}\right]_{T,\sigma}. \tag{11.90}$$

Comparison of Equations (11.89) and (11.90) identifies the partial derivatives in (11.89) with the various $\mu_i^{(s)}$ so that

$$Z^{(s)} = \sum_i n_i^{(s)} \mu_i^{(s)}. \tag{11.91}$$

The total differential of this expression is

$$dZ^{(s)} = \sum_i n_i^{(s)}\, d\mu_i^{(s)} + \sum_i \mu_i^{(s)}\, dn_i^{(s)}. \tag{11.92}$$

Upon equating the right sides of Equations (11.84) and (11.92) we get

$$\sum_i n_i^{(s)}\, d\mu_i^{(s)} + \Omega\, d\sigma + S^{(s)}\, dT = 0. \tag{11.93}$$

This is a Gibbs-Duhem relation (see Equation (8.71)) for the surface excess region.

If we divide Equation (11.93) by Ω, use Equation (11.55), and rearrange the equation, we obtain

$$d\sigma = -\frac{S^{(s)}}{\Omega}\,dT - \sum_i \Gamma_i^{(s)}\,d\mu_i^{(s)}. \tag{11.94}$$

This equation specifies the dependence of surface tension upon temperature and upon the chemical potentials of the various species in the system.

Consider a one-component system. The convention (11.57) which we have adopted for the dividing surface, applies. Since component 1 is the only species present, Equation (11.94), with Equation (11.56), becomes

$$-\frac{d\sigma}{dT} = \frac{S^{(s)}}{\Omega}. \tag{11.95}$$

Thus, the surface entropy may be determined from the temperature coefficient of the surface tension. Other relations may be derived, and the reader is referred to more exhaustive texts for this purpose. Here it should be pointed out that the excess entropy $S^{(s)}$, measured by the temperature coefficient of the surface tension, *is not any excess entropy, but only that referred to a dividing surface determined by the convention* (11.56). Thus, it may be seen that the convention may be used in a meaningful way even in the absence of detailed information concerning the structure of the interfacial layer.

As another example, consider a two-component system. If Equation (11.56) is employed, Equation (11.94) assumes the form

$$d\sigma = -\frac{S^{(s)}}{\Omega}\,dT - \Gamma_2^{(s)}\,d\mu_2^{(s)}. \tag{11.96}$$

If only isothermal changes are considered, Equation (11.94) simplifies to

$$-\left[\frac{\partial\sigma}{\partial\mu_2^{(s)}}\right]_T = \Gamma_2^{(s)}. \tag{11.97}$$

Thus, the surface concentration $\Gamma_2^{(s)}$ of the solute may be determined from the dependence of σ upon $\mu_2^{(s)}$. According to Equation (11.88), however, $\mu_2^{(s)}$ equals the chemical potential of the solute in either bulk phase. Thus, $\mu_2^{(s)}$ need not be measured directly. It will suffice to measure the chemical potential in the bulk phase.

In Equation (11.97) we are again confronted with a situation in which an excess quantity, this time a surface concentration, may be measured by means other than a detailed chemical analysis of the interfacial layer. *Once again, it is the surface density referred to the dividing surface prescribed by*

Equation (11.96) *which is measured* so that the convention has been employed in a useful manner even in the absence of detailed information concerning the structure of the interfacial layer.

14. The Gibbs Adsorption Isotherm

We may use the laws of ideal or dilute solution to further the discussion of Equation (11.96). The use may be illustrated in a simple manner by restricting consideration to a two-phase binary system. In view of Equation (11.88), $\mu_2^{(s)}$ may be replaced in Equation (11.97) by either $\mu_2^{(1)}$ or $\mu_2^{(2)}$, the chemical potentials of the solute in the bulk phases. By this device a facet of the problem is isolated to which Equation (9.1), delineated for a system with conventional variables, may be applied.

For definiteness, assume that the two phases 1 and 2 of the system consist of vapor and liquid with superscripts v and l, respectively, and replace $\mu_2^{(s)}$ by $\mu_2^{(v)}$, the chemical potential in the binary vapor. Then according to Equation (11.97),

$$\left[\frac{\partial \sigma}{\partial \mu_2^{(s)}}\right]_T = \left[\frac{\partial \sigma}{\partial \mu_2^{(v)}}\right]_T = -\Gamma_2^{(s)}, \tag{11.98}$$

where $\Gamma_2^{(s)}$ is referred to a dividing surface for which $\Gamma_1^{(s)}$ vanishes.

Substitution of Equation (9.1) into Equation (11.98) gives

$$\frac{1}{RT}\left[\frac{\partial \sigma}{\partial \ln x_2^{(v)}}\right]_{T,p} = -\Gamma_2^{(s)}. \tag{11.99}$$

This relation is known as the *Gibbs adsorption isotherm*. It relates the surface concentration $\Gamma_2^{(s)}$ to the concentration of solute in phase (v). Notice that the relation remains valid even when phase (l) is nonideal; only the ideality of phase (v) has been invoked.

If we pass to the special situation in which phase (v), the vapor phase, is pure component 2 and phase (l), the liquid, is pure component 1 (a not uncommon situation in many adsorption problems), Equation (9.3) may be employed for $\mu_2^{(v)}$ with the result

$$\mu_2^{(v)} = K(T) + RT \ln p, \tag{11.100}$$

and when this is substituted into Equation (11.98) we have

$$\frac{1}{RT}\left[\frac{\partial \sigma}{\partial \ln p}\right]_T = -\Gamma^{(s)}, \tag{11.101}$$

in which the subscripts have been dropped, since the gas phase consists of but a single component. This is another form of the Gibbs adsorption isotherm. Again, only the ideality of phase v is required.

XII

Systems in Gravitational and Centrifugal Fields

1. Application of the Generalized Feature of Equilibrium

The generalized feature of equilibrium, Equation (8.33), may be employed for the investigation of equilibrium in a system subject to the influence of an external field, for example gravitational or centrifugal.

It has been observed empirically that systems situated in gravitational or centrifugal fields develop inhomogeneities. The atmosphere furnishes an example of this, being dense at sea level and highly attenuated at great altitudes. These inhomogeneities are usually of the continuous sort and as a result, for the treatment of such systems, it will be necessary to define point density functions. We shall consider a system with the definite configuration illustrated in Figure 12.1. This consists of a fluid contained in a cylinder of cross section Λ whose axis lies along the direction of the external field. The direction of the field is indicated in the Figure. The z direction is also shown. It is antiparallel to the field direction. The length of the fluid in the z direction is h so that its volume V is

$$V = h\Lambda. \tag{12.1}$$

From symmetry the properties of the fluid should vary only in the z direction. If, for example, the internal energy per unit volume is defined as $\underline{u}$, then this function will depend upon z

$$\underline{u} = \underline{u}(z). \tag{12.2}$$

The internal energy of the volume $\Lambda\, dz$ in the interval dz will then be

$$\underline{u}(z)\Lambda\, dz. \tag{12.3}$$

Similarly, the entropy density may be defined as

$$\underline{s} = \underline{s}(z), \tag{12.4}$$

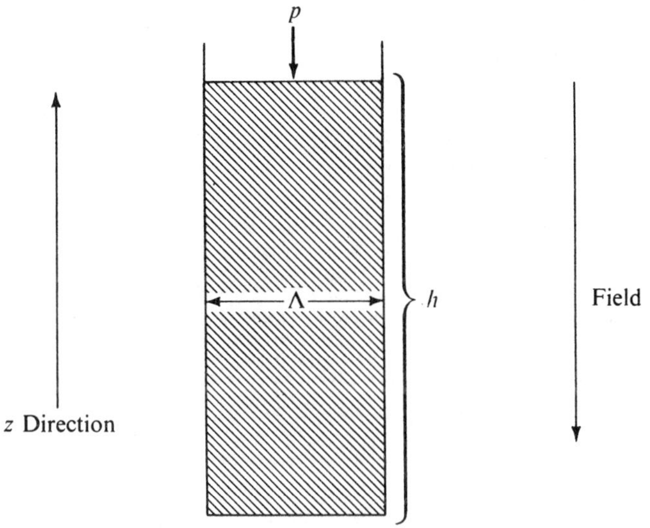

FIGURE 12.1. *A column of fluid in a gravitational field. The height of the column h and the cross sectional area Λ are shown.*

and the concentration of the *i*th species will be

$$c_i = c_i(z). \tag{12.5}$$

With these definitions, the total internal energy, entropy, and moles of the *i*th species in the system are given by the following integrals:

$$U = \int_0^h [\underline{u}\Lambda\, dz], \tag{12.6}$$

$$S = \int_0^h [\underline{s}\Lambda\, dz], \tag{12.7}$$

$$n_i = \int_0^h [c_i\Lambda\, dz]. \tag{12.8}$$

In the interest of notational simplicity, the z dependences of $\underline{u}$, $\underline{s}$, and c_i are not shown explicitly in these equations. Furthermore, the brackets enclosing the quantities under the integral sign, for the moment, signify nothing. They are introduced so that the forms may be consistent with a later usage.

The walls of the cylinder in Figure 12.1 are rigid so that all of the volume changes result in, and are measured by, a change in h,

$$dV = \Lambda\, dh. \tag{12.9}$$

As usual, the temperature is assumed to be uniform throughout the system. The pressure which the environment exerts upon the system at h is p; and therefore in this case,

$$Dw_{0_{\text{rev}}} = p\,dV = p\Lambda\,dh. \tag{12.10}$$

As usual, the precise form of $Dw_{0_{\text{rev}}}$ is important, and this is determined by the constraints χ_0 imposed on the system. The rigidity of the cylinder walls represents one of these constraints. The treatment and the choice of thermodynamic potential would have to be different if χ_0 were different. Thus, if the walls were not rigid, Equation (12.9) would not necessarily hold; whereas later Equation (12.9) makes possible a very convenient inversion of the order of integration and differentiation, provided that the most convenient thermodynamic potential is employed.

In view of Equation (12.10) the generalized feature of equilibrium (8.33) becomes

$$D\varphi = \delta U - T\,\delta S + p\,\delta V. \tag{12.11}$$

The differential of the Helmholtz function, $A = U - TS$, *imitates* this $D\varphi$ over a path of constant T and V (see Equation (8.25)). Thus, we may adopt Equation (8.31), $(\delta A)_{T,V} = 0$, as the extremal condition. Substitution of Equations (12.6) and (12.7) into Equation (8.31) yields

$$0 = (\delta A)_{T,V} = \left\{\delta\int_0^h [\underline{u}\Lambda\,dz]\right\}_{T,V} - T\left\{\delta\int_0^h [\underline{s}\Lambda\,dz]\right\}_{T,V} \tag{12.12}$$

Now according to Equation (12.9), if V is constant so is h.

(At this juncture, it is worth pointing out that had we used G rather than A as the thermodynamic potential—a perfectly valid method of proceeding—a path of constant T and p would have been involved. The counterpart of Equation (12.12) would then contain a $p\,\delta V$ term which would have been variable. The fact that h is constant enables one to bring the δ under the integral sign, and this simplifies the treatment. The analysis may be performed with G but it is more complicated. Since A and G (and several other potentials) are logically equivalent for the purpose at hand (see Chapter 8, Section 6), we might just as well use the one which allows the simplest procedure.)

2. Chemical Potential in an External Field

In accordance with the discussion of the last section, the δ's in Equation (12.12) may be brought under the integral sign with the result

$$\left\{\int_0^h \delta[\underline{u}\Lambda\,dz]\right\}_{T,V} - T\left\{\int_0^h \delta[\underline{s}\Lambda\,dz]\right\}_{T,V} = 0. \tag{12.13}$$

The reason for the brackets around $\underline{u}\Lambda\, dz$ and $\underline{s}\Lambda\, dz$ is now evident. We wish to distinguish clearly the virtual variation δ from the differential of height dz.

Now the volume $\Lambda\, dz$ may be regarded as a thermodynamic subsystem—a phase of thickness dz located at z (we recall the definition of the thermodynamic system given in Chapter 1, Section 3). Its volume is constant at $\Lambda\, dz$ by definition, and so it cannot perform work upon its surroundings. The number of moles of the ith species contained within this phase is obviously (see Equation (12.8))

$$c_i\Lambda\, dz, \tag{12.14}$$

and the internal energy is given by Equation (12.3) while the entropy is

$$s\Lambda\, dz. \tag{12.15}$$

In analogy to Equation (7.3) we may therefore write

$$\delta[\underline{u}\Lambda\, dz] = T\delta[\underline{s}\Lambda\, dz] + \sum_i \mu_i(z)\, \delta[c_i\Lambda\, dz], \tag{12.16}$$

where

$$\mu_i(z) = \left\{\frac{\partial[\underline{u}\Lambda\, dz]}{\partial[c_i\Lambda\, dz]}\right\}_{[\underline{s}\Lambda\, dz]\,[\Lambda\, dz]}, \tag{12.17}$$

and in which, unlike Equation (7.3), no volume differential term appears because $\Lambda\, dz$ is constant. It is clear that $\mu_i(z)$ is the usual chemical potential in the sense that it is the derivative of the internal energy with respect to mole number; temperature, volume, and the mole numbers of other species being held constant (see Equation (8.43)). The last term in Equation (12.16) appears because, although the entire system is *closed*, the individual phases of thicknesses dz are *open*.

Substitution of Equation (12.17) into Equation (12.13) yields

$$\int_0^h \sum_i \mu_i(z)\, \delta[c_i\Lambda\, dz] = 0. \tag{12.18}$$

Since the total system is closed, we have, from Equation (12.8),

$$0 = \int_0^h \delta[c_i\Lambda\, dz]. \tag{12.19}$$

Equations (12.18) and (12.19) are exactly the same as Equations (8.60) and (8.61) except for the fact that in both equations the summation over j has been replaced by an integration over z. Thus, we must arrive at the result (8.63) which in the present instance takes the form

$$\mu_i(z) = \lambda_i, \tag{12.20}$$

where λ_i is independent of z. Thus, even in the presence of a field, the condition of internal equilibrium is that the chemical potential shall be uniform throughout the system.

3. Role of the External Field

At this point, the reader may begin to wonder what the external field has to do with the problem, for the condition of internal equilibrium (12.20) has actually been derived without the explicit appearance of any quantities associated with the field. Our only concession to its presence has been the admission that the system might be continuously inhomogeneous.

This state of affairs illuminates a subtle point which seems to have been overlooked by the authors of many texts on thermodynamics. The point is less likely to be overlooked when a careful analysis of the relation between *work* and the first law (of the sort presented in Chapter 3) is performed, and when derivations are based upon Equation (8.33) rather than upon analogy. Many authors advance the following argument.

Consider the field of Figure 12.1 and assume that the potential energy of a mass m at the height z in this field is

$$\xi(z) = mk_0z. \tag{12.21}$$

If a gravitational field is involved,

$$k_0 = g, \tag{12.22}$$

where g is the acceleration of gravity, whereas if the field is centrifugal,

$$k_0 = -\tfrac{1}{2}\omega^2 z, \tag{12.23}$$

where ω is the angular velocity (radians per second). Now the work expended in moving a simple mass from the reference height to z is just $\xi(z)$. It is assumed by many that the same reasoning may be applied to the transfer of δn_i moles of species i in the system of Figure 12.1 from the bottom of the system to the height z. Thus, according to this point of view, if the molecular weight of the species is M_i, the mass involved will be $M_i\,\delta n_i$, and by Equation (12.21) the reversible work involved in the transfer will be

$$Dw = (M_i\,\delta n_i)k_0z. \tag{12.24}$$

Actually it is difficult, if not impossible, to conceive of a method for effecting this transfer experimentally *within* the system of Figure 12.1 while at the same time measuring the reversible work. All conceivable methods always involve other processes (for example, the movement of other species) so that the work in question cannot be isolated. Thus, there is no direct macroscopic method for determining the *form* of the field work in terms of changes observable in the environment. As a result, it is best not to introduce the field into the thermodynamic treatment through the medium of *work*. This is why field terms do not appear in Equations (12.10) and (12.11), and why the field quantities do not enter explicitly into the proof of Equation (12.20).

On the other hand, we can introduce the field through the definition of $\mu_i(z)$ in Equation (12.17). As a matter of fact, since we are concerned with local properties, it is simpler to consider a uniform phase located at the elevation z. At this position the change in internal energy of the phase which accompanies the addition of a mass consisting of δn_i moles of species i must be due partly to the fact that the added mass has potential energy. Thus,

$$dU = dU_0 + d\xi_i \tag{12.25}$$

and

$$\mu_i = \left[\frac{\partial U}{\partial n_i}\right]_{S,V} = \left[\frac{\partial U_0}{\partial n_i}\right]_{S,V} + \left[\frac{\partial \xi_i}{\partial n_i}\right]_{S,V}, \tag{12.26}$$

where dU_0 is that part of the internal energy change not due to the change in potential energy.

Clearly, in accordance with Equation (12.21),

$$d\xi_i = (M_i\, dn_i)k_0 z, \tag{12.27}$$

so that

$$\mu_i = \left[\frac{\partial U_0}{\partial n_i}\right]_{S,V} + M_i k_0 z. \tag{12.28}$$

If the field vanishes ($k_0 = 0$),

$$\mu_i = \left[\frac{\partial U_0}{\partial n_i}\right]_{S,V}, \tag{12.29}$$

and the derivative on the right is the ordinary field-free chemical potential. The fundamental question is the following. Is dU_0 independent of k_0, that is, is it independent of the field? Are the two terms in Equation (12.28) entirely uncoupled? The assumption which is made (and it must be recognized for what it is, an *assumption*) is that the terms are independent of one another, and Equation (12.28) is usually expressed in the form

$$\mu_i(z) = \mu_i^*(z) + M_i k_0 z, \tag{12.30}$$

where $\mu_i^*(z)$ is the chemical potential measured in the absence of the field, all other things remaining the same.

Equation (12.30) is therefore not derived from the fundamental principles of thermodynamics. It is in the same class as the assumptions concerning the choice of variables of state. It is justified by the degree of correspondence between the theoretical predictions which can be derived from it and the results of experiments. This point is not always made clear.

4. The Barometric Formula

We now illustrate the combined use of Equations (12.30) and (12.20) by treating a particularly simple system. This is the case in which the fluid in the cylinder of Figure 12.1 is a one-component ideal gas.

In this case, with variations in temperature ruled out, Equation (9.9) becomes

$$d\mu^* = \frac{V}{n}\,dp, \tag{12.31}$$

where μ^* is the chemical potential defined in the absence of a field (since Equation (9.9) is derived in the absence of fields). Now in the system of Figure 12.1, the pressure p will depend upon z so that μ^* will also depend on z by way of Equation (12.31). Thus, we may write

$$\frac{d\mu^*}{dz} = \frac{V}{n}\frac{dp}{dz}. \tag{12.32}$$

On the other hand, in the presence of the field the appropriate chemical potential is given by μ of Equation (12.30), and by substituting Equation (12.30) into Equation (12.32) we obtain

$$\frac{d\mu^*}{dz} = \frac{V}{n}\frac{dp}{dz} = \frac{d\mu}{dz} - M\frac{d}{dz}(k_0 z). \tag{12.33}$$

But at equilibrium Equation (12.20) requires

$$\frac{d\mu}{dz} = 0, \tag{12.34}$$

so that Equation (12.33) becomes

$$\frac{V}{n}\frac{dp}{dz} = -M\frac{d}{dz}(k_0 z). \tag{12.35}$$

If the field is gravitational, Equation (12.22) may be employed, and for $\frac{V}{n}$ for an ideal gas, we derive from Equation (5.1)

$$\frac{V}{n} = \frac{RT}{p}. \tag{12.36}$$

Thus, Equation (12.35) becomes

$$RT\frac{d\ln p}{dz} = -Mg, \tag{12.37}$$

which may be integrated immediately to yield

$$p = p_0 e^{-\frac{Mgz}{RT}}, \tag{12.38}$$

where p_0 is the pressure at $z = 0$. This result is the famous *barometric formula* which describes, approximately, the isothermal attenuation of the atmosphere with increasing altitude z.

5. Centrifugal Fields

The laws of ideal solution may be used to illustrate the case of a binary solution in a centrifugal field. In this case, Equation (12.30) allows us to extract a feature of the system to which Equation (9.1) may be applied, since μ_i^* is by definition the function specified by Equation (9.1) when the solution is ideal.

If Equations (12.23) and (9.1) are substituted into Equation (12.30) and the result is differentiated with respect to z, we obtain (since at equilibrium μ_i is independent of z),

$$\left[\frac{\partial \mu_i}{\partial z}\right]_T = 0 = \left[\frac{\partial \mu_i^0}{\partial z}\right]_T + RT\left[\frac{\partial \ln x_i}{\partial z}\right]_T - M_i\omega^2 z. \qquad (12.39)$$

Now

$$\left[\frac{\partial \mu_i^0}{\partial z}\right]_T = \left[\frac{\partial \mu_i^0}{\partial p}\right]_T\left[\frac{\partial p}{\partial z}\right]_T = V_i^0\left[\frac{\partial p}{\partial z}\right]_T, \qquad (12.40)$$

where Equation (5.66) has been used. Substitution of this into Equation (12.39) gives

$$V_i^0\left[\frac{\partial p}{\partial z}\right]_T + RT\left[\frac{\partial \ln x_i}{\partial z}\right]_T = M_i\omega^2 z. \qquad (12.41)$$

If Equation (12.41) is multiplied by x_i and summed over i, the result is

$$\left[\frac{\partial p}{\partial z}\right]_T \sum_i x_i V_i^0 + RT\sum_i\left[\frac{\partial x_i}{\partial z}\right]_T = \omega^2 z \sum_i x_i M_i. \qquad (12.42)$$

Since $\sum_i x_i = 1$, it follows that the second sum on the left of Equation (12.42) is zero. Thus, we arrive at the result

$$\left[\frac{\partial p}{\partial z}\right]_T = \frac{\omega^2 z \sum_i x_i M_i}{\sum_i x_i V_i^0}. \qquad (12.43)$$

This equation may be substituted into (12.41) to yield

$$\frac{V_i^0\omega^2 z \sum_j x_j M_j}{\sum_j x_j V_j^0} + RT\left[\frac{\partial \ln x_i}{dz}\right]_T = M_i\omega^2 z. \qquad (12.44)$$

In the simple case of a binary solution Equation (12.44) becomes, for component 1,

$$\frac{V_1^0\omega^2 z(x_1M_1 + x_2M_2)}{x_1V_1^0 + x_2V_2^0} + RT\frac{\partial \ln x_1}{\partial z} = M_1\omega^2 z, \tag{12.45}$$

with a similar equation for component 2. Thus, the mole fraction x_1 of component 1 at point z will depend on M_1 and M_2 and will differ from the dependence of x_2 upon z. This behavior forms the basis of the separation of large molecules by means of a centrifuge. It also provides a method for the determination of the molecular weights of macromolecules.

XIII

Elastic Systems

1. Stress and Strain

Thermodynamics may also be applied to problems of equilibrium in elastic solids. The theory of small elastic deformations is an elegant and detailed subject, and we have no intention of treating it at length in this book.† On the other hand, elastic constraints make possible the performance of a new kind of work, and it is of interest to see how this new class of work modifies the appropriate thermodynamic and chemical potentials.

The study of elasticity is facilitated by the introduction of the formal concepts of strain and stress. In the general treatment of the subject, the strain and stress fields are each fully characterized by six tensorial components, each of which may be dependent upon position within the system. For the case of small deformations, this formulation is equal to the task of dealing with both isotropic and anisotropic systems possessing varying degrees of symmetry. From the point of view of acquiring insight into *how* thermodynamics may be applied to elasticity, not much is gained from a study of complicated systems, and we shall confine our attention to isotropic systems subjected to very simple conditions of stress.

Ordinarily the *strain* $\mathbf{e}$ at a point in an elastic body is defined as the fractional change in vector displacement, $\mathbf{dr}$, of two points originally separated by the distances $\mathbf{ds}$ in the undeformed body. Thus,

$$\mathbf{e} = \frac{\mathbf{dr}}{|\mathbf{ds}|}. \tag{13.1}$$

The various displacements are illustrated in Figure 13.1. Initially in the undeformed solid two neighboring points P and Q are separated by the vector distance $\mathbf{ds}$. After deformation, the relative separation is $\mathbf{ds}'$ and the point Q is now denoted by Q'. In general, P is also displaced to a new point

† For a more adequate but still not exhaustive treatment of the formalism of small deformations, see H. B. Callen, *Thermodynamics* (John Wiley, 1960), Chapter 13.

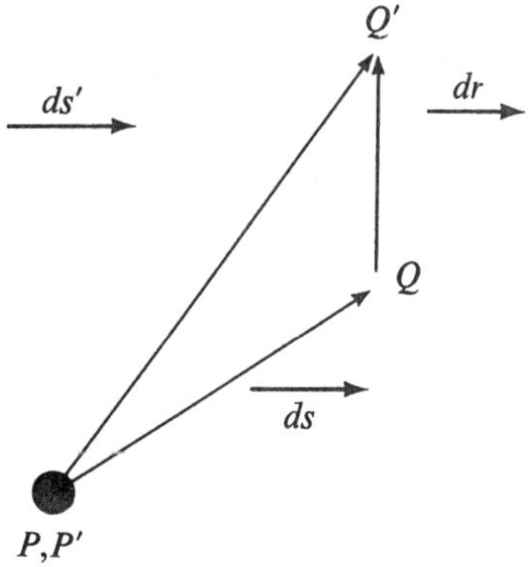

FIGURE 13.1. *Relative displacement of two points P and Q in a deformed body when the local deformation carries P to P'. Under this condition, Q goes to Q' so that* **dr** *represents the shift of Q relative to P. The local deformation is not shown, in the interest of simplicity, and P' in the figure has been made to coincide with P.*

P'; but since we are only interested in the relative displacement of the two points P and Q, this displacement is not shown in the figure.

We must also define *stress*. Consider an arbitrary plane surface within a solid. If the solid is strained the material on one side of this surface will exert forces on the material on the other side and vice versa. These forces will have a definite direction which, among other things, will depend upon the orientation of the surface. If the area of the surface is $d\Omega$, then the force **df** acting on $d\Omega$ will be

$$\mathbf{df} = \boldsymbol{\tau}\, d\Omega \tag{13.2}$$

where $\boldsymbol{\tau}$ is the *stress* across the surface. Thus, *stress* is the force per unit area acting across a given surface.

2. Thermodynamic Potential of a Simple Isotropic System

As we have indicated, only stress-strain systems of the simplest symmetry will be treated here. Consider the system of Figure 13.2. In the figure the rectangular shape is an isotropic solid solution (phase 1) which may contain a number of components. This phase is surrounded by a fluid solution (phase 2) composed of the same components. The solid is stressed by a uniform pressure p in the fluid phase 2 and by a tensile stress τ (arrow in figure) which acts in the z direction. Since the pressure p is a hydrostatic stress, it acts normal to the solid surface. Thus, the actual stress in the z direction is

$$\tau_z = \tau - p. \tag{13.3}$$

The stress on the vertical sides of the solid (in the x and y directions) will be

$$\tau_x = \tau_y = -p. \tag{13.4}$$

Assume that the length of the solid with $\tau = 0$ is L and that for simplicity the cross section is square with side l. The area of the cross section is then l^2. With this simple set of stresses the isotropic solid will deform in a simple manner. The length L will be extended to $L + dL$ and the thickness l will be changed to $l - dl$ (with dl positive). Symmetry demands that the strain (see Equation (13.1)) be uniform throughout the solid. The component of this uniform strain in the z direction will be dL/L and in the x and y directions it will be $-dl/l$.

The work performed *on* the solid during this deformation may be computed as follows. The area of a vertical side is Ll and stress is force per unit area so that the force on a vertical side is $-pLl$. This force is normal to the side which moves the distance $-dl$ during the deformation. Thus, the work on one side is $pLl\,dl$. Since two displacements must be considered, one in the x and the other in the y direction, the total work on the sides is $2pLl\,dl$. In a similar manner, the force on the horizontal side which is displaced by dL in the z direction is $(\tau - p)l^2$ and the work is $(\tau - p)l^2\,dL$. The total work performed *on* the solid is therefore

$$Dw^{(1)}_{0_{\text{rev}}} = (\tau - p)l^2\,dL + 2pLl\,dl. \tag{13.5}$$

The work $Dw^{(2)}_{0_{\text{rev}}}$ performed *by* the fluid phase 2 is as usual

$$Dw^{(2)}_{0_{\text{rev}}} = p\,dV^{(2)}. \tag{13.6}$$

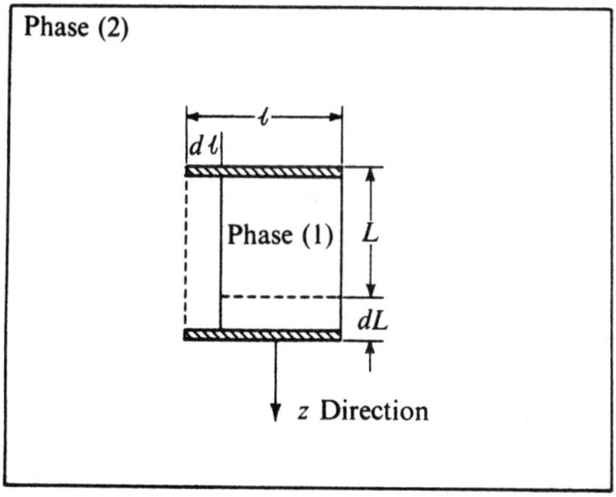

FIGURE 13.2. *Deformable isotropic solid solution, phase* (1) *surrounded by a fluid solution, phase* (2).

Now the increment of $dV^{(1)}$ in the volume of phase 1 is

$$dV^{(1)} = -2Ll\,dl + l^2\,dL. \tag{13.7}$$

Substitution of this into Equation (13.5) yields

$$Dw^{(1)}_{0_{\text{rev}}} = p\,dV^{(1)} - \tau l^2\,dL. \tag{13.8}$$

Thus, the total value of $Dw_{0_{\text{rev}}}$ to be used in Equation (8.33) is given by

$$Dw_{0_{\text{rev}}} = Dw^{(1)}_{0_{\text{rev}}} + Dw^{(2)}_{0_{\text{rev}}} = p\,dV^{(1)} + p\,dV^{(2)}$$

$$- \tau l^2 dL = p\,dV - \tau l^2 dL, \tag{13.9}$$

where

$$V = V^{(1)} + V^{(2)} \tag{13.10}$$

is the total volume of the system. According to Equation (8.3) then,

$$D\varphi = \delta U + p\,\delta V - \tau l^2\,\delta L - T\,\delta S = 0. \tag{13.11}$$

It is possible to synthesize many functions which can serve as thermodynamic potentials. Consider

$$\Xi = U + pV - \tau l^2 L - TS. \tag{13.12}$$

Along a path of constant T, p, and τl^2 this function will *imitate* $D\varphi$. Thus,

$$(\delta\Xi)_{T,p,\tau l^2} = \delta U + p\,\delta V - \tau l^2\,\delta L - T\,\delta S, \tag{13.13}$$

$$(D\varphi)_{T,p,\tau l^2} = \delta U + p\,\delta V - \tau l^2\,\delta L - T\,\delta S, \tag{13.14}$$

and the right sides of these equations are identical. Therefore, Ξ is a suitable thermodynamic potential for this system, and the extremal condition becomes

$$(\delta\Xi)_{T,p,\tau l^2} = 0. \tag{13.15}$$

3. The Chemical Potential

In the usual manner, U and S may be partitioned between the phases

$$U = U^{(1)} + U^{(2)}, \tag{13.16}$$

$$S = S^{(1)} + S^{(2)}, \tag{13.17}$$

so that Ξ assumes the form

$$\Xi = [U^{(1)} + pV^{(1)} - (\tau l^2)L - TS^{(1)}] + (U^{(2)} + pV^{(2)} - TS^{(2)})$$

$$= M^{(1)} + G^{(2)}, \tag{13.18}$$

where $G^{(2)}$ is a Gibbs function but $M^{(1)}$ is not. This expression may be substituted into Equation (13.14)

$$(\delta\Xi)_{T,p,\tau l^2} = (\delta M^{(1)})_{T,p,\tau l^2} + (\delta G^{(2)})_{T,p} = 0. \tag{13.19}$$

As usual,

$$dU^{(2)} = T\,dS^{(2)} - p\,dV^{(2)} + \sum_i \mu_i^{(2)}\,dn_i^{(2)}, \tag{13.20}$$

where $\mu_i^{(2)}$ is the usual chemical potential. For phase 1,

$$dU^{(1)} = T\,dS^{(1)} - p\,dV^{(1)} - (\tau l^2)\,dL + \sum_i \mu_i^{(1)}\,dn_i^{(1)}, \tag{13.21}$$

where

$$\mu_i^{(1)} = \left[\frac{\partial U^{(1)}}{\partial n_i^{(1)}}\right]_{S,V,L}. \tag{13.22}$$

The total differentials of $M^{(1)}$ and $G^{(2)}$ are

$$dM^{(1)} = dU^{(1)} + p\,dV^{(1)} + V^{(1)}\,dp - (\tau l^2)\,dL - Ld(\tau l^2) - T\,dS^{(1)} - S^{(1)}\,dT, \tag{13.23}$$

and

$$dG^{(2)} = dU^{(2)} + p\,dV^{(2)} + V^{(2)}\,dp - T\,dS^{(2)} - S^{(2)}\,dT. \tag{13.24}$$

Substitution of Equations (13.20) and (13.21) into Equations (13.23) and (13.24) gives

$$dM^{(1)} = V^{(1)}\,dp - Ld(\tau l^2) - S^{(1)}\,dT + \sum_i \mu_i^{(1)}\,dn_i^{(1)} \tag{13.25}$$

and

$$dG^{(2)} = V^{(2)}\,dp - S^{(2)}\,dT + \sum_i \mu_i^{(2)}\,dn_i^{(2)}. \tag{13.26}$$

From Equation (13.25) it is apparent that another definition of $\mu_i^{(2)}$, logically equivalent to Equation (13.22), is

$$\mu_i^{(1)} = \left[\frac{\partial M^{(1)}}{\partial n_i^{(1)}}\right]_{T,p,\tau l^2}. \tag{13.27}$$

Substitution of Equations (13.25) and (13.26) into Equation (13.19) yields

$$\sum \mu_i^{(2)}\,\delta n_i^{(2)} + \sum_i \mu_i^{(1)}\,\delta n_i^{(1)} = 0, \tag{13.28}$$

which together with the conservation conditions

$$\delta n_i^{(2)} = -\delta n_i^{(1)}, \qquad i = 1, 2, \ldots, C \tag{13.29}$$

requires as a condition of internal equilibrium the usual result

$$\mu_i^{(1)} = \mu_i^{(2)}. \tag{13.30}$$

4. Vapor Pressure of an Elastic System

Consider the special case in which the elastic phase (phase 1) of Figure 13.2 consists of only one component, while phase 2 is the vapor of this component. The condition of equilibrium (Equation (13.30)) simplifies to

$$\mu^{(1)} = \mu^{(2)}; \tag{13.31}$$

and for the persistence of equilibrium, we have, in analogy to Equation (9.8),

$$d\mu^{(1)} = d\mu^{(2)}. \tag{13.32}$$

For phase 2 we may borrow from Equation (9.10) and write

$$d\mu^{(2)} = v^{(2)0}\,dp - s^{(2)0}\,dT, \tag{13.33}$$

where $v^{(2)0}$ and $s^{(2)0}$ are the volume and entropy per mole.

By employing the definition (13.27) and inverting the order of differentiation as in Equation (5.65), it is an easy matter to develop the counterpart of Equation (9.10) for the elastic phase. Thus, by a procedure similar to that which led to Equation (9.10) we obtain

$$d\mu^{(1)} = v^{(1)0}\,dp - s^{(1)0}\,dT - \frac{L}{n^{(1)}}\,d(\tau l^2), \tag{13.34}$$

where

$$v^{(1)0} = \bar{v}^{(1)} = \left[\frac{\partial V}{\partial n^{(1)}}\right]_{T\ p,\tau l^2}, \tag{13.35}$$

with a similar expression for the $s^{(1)0} = \bar{s}^{(1)}$. Furthermore,

$$\frac{L}{n^{(1)}} = L = \left[\frac{\partial L}{\partial n^{(1)}}\right]_{T,p,\tau l^2}. \tag{13.36}$$

The variables $v^{(1)}$, $\bar{s}^{(1)}$, and L are each functions of T, p, and τl^2.

Now introduction of Equations (13.33) and (13.34) into Equation (13.32) yields

$$(v^{(1)0} - v^{(2)0})\,dp - (s^{(1)0} - s^{(2)0})\,dT = \frac{L}{n^{(1)}}\,d(\tau l^2). \tag{13.37}$$

Suppose we restrict our attention to the constant pressure case, $dp = 0$, and apply Equation (9.15). Under these conditions, Equations (13.37) becomes

$$\left[\frac{\partial(\tau l^2)}{\partial T}\right] = -\frac{n^{(1)}\lambda_v}{TL}, \tag{13.38}$$

where λ_v is the heat of vaporization. This is the stress analogue of the Clapeyron equation (9.16).

Another relation is obtained if T is held constant so that $dT = 0$. This is,

$$\left[\frac{\partial p}{\partial(\tau l^2)}\right] = \frac{L}{n^{(1)}(v^{(1)0} - v^{(2)0})}. \tag{13.39}$$

Since phase 2 is a gas, we may assume that

$$v^{(2)0} \gg v^{(1)0}, \tag{13.40}$$

and if the gas is assumed to be ideal,

$$V^{(2)0} = \frac{RT}{p}. \tag{13.41}$$

Substitution of Equations (13.40) and (13.41) into Equation (13.39) yields

$$\left[\frac{\partial \ln p}{\partial(\tau l^2)}\right]_T = -\frac{L}{n^{(1)}RT}. \tag{13.42}$$

This equation determines the dependence of vapor pressure upon tensile force τl^2.

Since the right-hand side of Equation (13.42) is negative, an increase in tensile force τl^2 decreases the vapor pressure.

For an excellent treatment of other aspects of the thermodynamics of elastic systems, the reader is referred to Chapter 13 of H. B. Callen's book (see Reference 3).

XIV

Stability

1. Stable Equilibrium

The inequality, $\Delta\varphi > 0$, derived in Chapter 8, was qualified by several conditions. The argument in Section 4 of that chapter is founded upon the *assumption* that the initial state from which the variation, corresponding to the inequality, proceeds is one of *equilibrium in the following sense.* If a system is in equilibrium subject to a set of constraints $\chi_0 + \chi_1$ and χ_1 is removed, then the initial state is the one to which the system always goes as long as χ_0 is retained. This means that if the system is in the initial state and is displaced from this state by a "variational" constraint χ_1, it will always return to it when χ_1 is removed. A careful examination of the argument advanced in Section 4 of Chapter 8 will reveal that the properties of this behavior were used in arriving at the result $\Delta\varphi > 0$.

The property of returning to the initial state, when the variational constraint is withdrawn, identifies the initial state as one of "stable" equilibrium subject to the constraints χ_0. The property of return is actually used in order to define "stable" equilibrium. Hence, $\Delta\varphi > 0$ is a feature of stable equilibrium.

Now the definition of equilibrium offered in Chapter 1, Section 3 provides no means of distinguishing stable equilibrium (since it only requires that a system exist in a reproducible and time-independent condition), and therefore applies to a more general class of equilibria. By the same token, the elementary method presented in Chapter 8, Section 2 does not distinguish stable equilibria and is therefore applicable to the entire class delineated by the definition in Chapter 1.

The same may be said of the generalized "feature" of equilibrium represented by the extremal condition, Equation (8.33). It does not distinguish stable equilibrium. For example, (8.33) applies equally well if the extremum is a minimum or a maximum, whereas $\Delta\varphi > 0$ implies that the initial state is always the site of a minimum. Thus, "stable" equilibrium may also be characterized by the fact that the function of state, whose increment imitates $\Delta\varphi$, has a minimum in the initial state.

This requirement of minimization may be employed to deduce certain disallowed functional relationships among the variables of state of a system presumed stable. We shall give some examples in the present chapter.

2. Mechanical Stability

As in all other considerations in this book, we emphasize the fact that relationships are true only in the light of the particular variables, constraints, and kinds of work with which a system is associated. Thus, a system which is unstable subject to a given set of constraints χ_0 may be made stable by the introduction of another constraint. Furthermore a system which is stable when the constraints χ_0 have numerical values which lie in one range may become unstable when these values enter another range. The discussion surrounding the S-shaped isotherm of Figure 14.2 is an example of this separation into several kinds of ranges.

The discussion will be confined to the simplest kind of systems. Once again, the focus will be on method rather than detail. Thus, consider a simple one-component, single-phase fluid whose constraints χ_0 are such that it may only perform volume work. Then the Helmholtz free energy A may be used as a thermodynamic potential (see Chapter 8, Section 5). If the fluid is confined to a rigid container of volume V as in Figure 14.1, and its temperature is maintained at T by a thermostat, then variations over a path of constant T and V may be carried out by inserting a partition, as in the figure, in order to divide the system into two equal halves and to compress one while expanding the other. The partition, of course, represents a constraint χ_1. If the fluid is initially in stable equilibrium, it follows, from Chapter 8, Section 5, that

$$(\Delta A)_{T,V} = (\Delta \varphi)_{T,V} > 0. \tag{14.1}$$

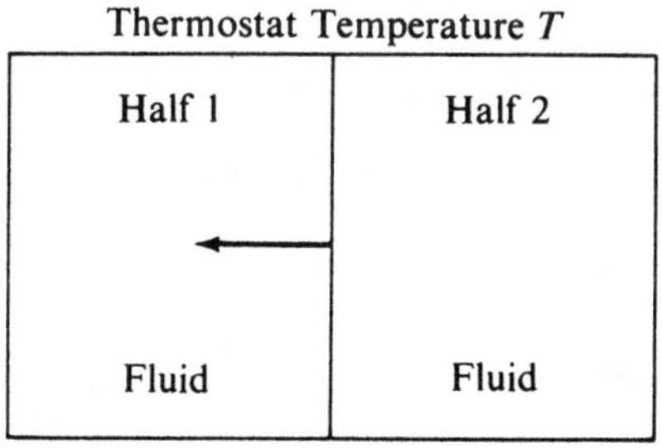

FIGURE 14.1. *One-component single-phase fluid confined to a thermostatted container of fixed volume V. The container is divided into two equal parts by a partition which is used to compress half 1 while allowing half 2 to expand.*

We may now designate thermodynamic quantities which refer to half 1 by appending the subscript 1, and those belonging to half 2 by the subscript 2. Thus,

$$A = A_1 + A_2, \tag{14.2}$$

$$V = V_1 + V_2. \tag{14.3}$$

During the variation of Figure 14.1, it is clear, since V is fixed, that

$$\Delta V_2 = -\Delta V_1. \tag{14.4}$$

The states of the halves of the fluid are determined by the independent variables T, V_1, and T, V_2, respectively. Furthermore, it is clear that initially,

$$V_1 = V_2 = \frac{V}{2} \tag{14.5}$$

and

$$A_1 = A_2 = \frac{A}{2}. \tag{14.6}$$

Expansion of A_1 and A_2 in Taylor's series in the variations ΔV_1 and ΔV_2 gives

$$A_1 = A_1\left(\frac{V}{2}\right) + \left[\frac{\partial A_1\left(\frac{V}{2}\right)}{\partial V_1}\right]_{T,n/2} \Delta V_1 + \frac{1}{2}\left[\frac{\partial^2 A_1\left(\frac{V}{2}\right)}{\partial V_1^2}\right]_{T,n/2} (\Delta V_1)^2 + \cdots \tag{14.7}$$

and

$$A_2 = A_2\left(\frac{V}{2}\right) + \left[\frac{\partial A_2\left(\frac{V}{2}\right)}{\partial V_2}\right]_{T,n/2} \Delta V_2 + \frac{1}{2}\left[\frac{\partial^2 A_2\left(\frac{V}{2}\right)}{\partial V_2^2}\right]_{T,n/2} (\Delta V_2)^2 + \cdots, \tag{14.8}$$

where n is the number of moles in the entire system and $\frac{n}{2}$ the number in each half. By adding Equations (14.7) and (14.8) we get

$$\left[A_1 - A_1\left(\frac{V}{2}\right)\right] + \left[A_2 - A_2\left(\frac{V}{2}\right)\right] = \left[\Delta A_1 + \Delta A_2\right]_{T,V} = (\Delta A)_{T,V}$$

$$= \left\{\left[\frac{\partial A_1\left(\frac{V}{2}\right)}{\partial V_1}\right]_{T,n/2} - \left[\frac{\partial A_2\left(\frac{V}{2}\right)}{\partial V_2}\right]_{T,n/2}\right\} \Delta V_1$$

$$+ \frac{1}{2}\left\{\left[\frac{\partial^2 A_1\left(\frac{V}{2}\right)}{\partial V_1^2}\right]_{T,n/2} + \left[\frac{\partial^2 A_2\left(\frac{V}{2}\right)}{\partial V_2^2}\right]_{T,n/2}\right\} (\Delta V_1)^2 + \cdots, \tag{14.9}$$

where Equations (14.2) and (14.4) have been used. Since both A and V are proportional to n, the number of moles in the system (intensive variables remaining constant), it follows that

$$\left[\frac{\partial A_1\left(\frac{V}{2}\right)}{\partial V_1}\right]_{T,n/2} = \left[\frac{\partial A_2\left(\frac{V}{2}\right)}{\partial V_2}\right]_{T,n/2} = \left[\frac{\partial A(V)}{\partial V}\right]_{T,n} \tag{14.10}$$

and

$$\left[\frac{\partial^2 A_1\left(\frac{V}{2}\right)}{\partial V_1^2}\right]_{T,n/2} = \left[\frac{\partial^2 A_2\left(\frac{V}{2}\right)}{\partial V_2^2}\right]_{T,n/2} = \left[\frac{\partial^2 A(V)}{\partial V^2}\right]_{T,n} = -\left[\frac{\partial p}{\partial V}\right]_{T,n}, \tag{14.11}$$

where in Equation (14.11) we have used Equation (4.85).

Substitution of Equations (14.10) and (14.11) into the last equation of (14.9) and the result into Equation (14.1) yields

$$(\Delta A)_{T,V} = -\left[\frac{\partial p}{\partial V}\right]_{T,n} (\Delta V_1)^2 + \cdots > 0. \tag{14.12}$$

For a small enough variation, terms beyond the second order may be ignored, and since $(\Delta V_1)^2$ is always positive Equation (14.12) reduces to

$$\left[\frac{\partial p}{\partial V}\right]_{T,n} < 0. \tag{14.13}$$

This is frequently referred to as the condition of mechanical stability. It implies that the isothermal compressibility β (Equation (6.2)) is always positive when the system is stable.

The reader must be cautioned about the too-general interpretion of Equation (14.13). It is a necessary *consequence* of the assumption that the system is initially in a state of stable equilibrium; not a proof. Furthermore, it applies to a system whose initial constraints χ_0 are just those which have been assigned. There is no reason why, in principle, the set χ_0 could not be augmented so that even when Equation (14.13) is violated the system remains stable. Of course, changing the set of initial constraints χ_0 will mean that A can no longer be employed as the appropriate thermodynamic potential; therefore, Equations (14.1) and (14.13), which are derived from it, are no longer valid for stability.

A familiar situation to which Equation (14.13) has been applied is the Van der Waals fluid[16] whose $p - V$ isotherm resembles the S-shaped curve

[16] P. S. Epstein, *Textbook of Thermodynamics* (New York: John Wiley and Sons, Inc., 1937), p. 10.

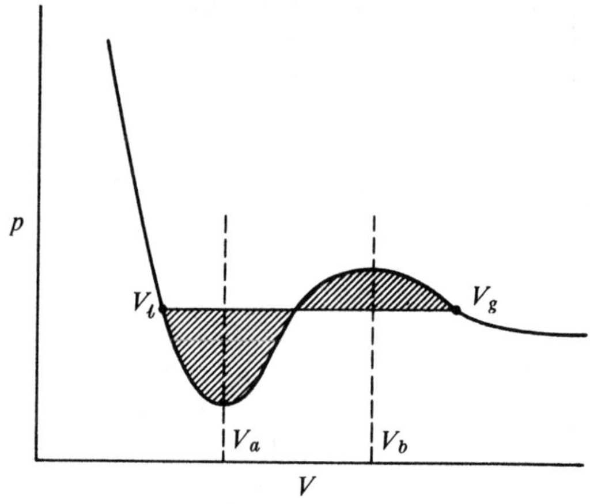

FIGURE 14.2. *Isotherm for one mole of Van der Waals fluid. Volumes lying between V_a and V_b are in the unstable region.*

in Figure 14.2. When the volume lies in the range between V_a and V_b in the figure,

$$\left[\frac{\partial p}{\partial V}\right]_{T,n} > 0, \tag{14.14}$$

which clearly violates Equation (14.13). Thus, in this region the single phase fluid is unstable and must separate into two phases of molar volumes V_l and V_v (liquid and vapor), respectively. If there are no constraints causing either superheating of the liquid or supersaturation of the vapor, the actual isotherm will include the horizontal line, as shown, with the shaded loops having equal areas. The regions between V_l and V_a and V_b and V_v are regions of superheating and supersaturation, respectively, and in many cases have been partially realized experimentally.

The Van der Waals isotherm furnishes an example of how the same constraints χ_0 can have ranges (for their numerical values) of both stability and instability. Also, we remind the reader that it may be possible to add some very generalized "thought" constraint which would render the range between V_a and V_b stable in spite of Equation (14.14). However, in this case A would no longer be a suitable thermodynamic potential and Equation (14.13) would no longer be a requirement for stability.

The "conditions of stability", like the "features" of equilibrium depend upon the variables, constraints, and kinds of work associated with the system.

3. Thermal Stability

Many relations similar to Equation (14.13) may be derived by employing essentially the same technique. Thus, if we continue to deal with a one-component, single-phase system constrained so as to perform volume work only, the enthalpy may be used as a thermodynamic potential as well as the Helmholtz free energy. Thus, instead of Equation (14.1) we might use (see Chapter 8, Section 5)

$$(\Delta H)_{S,p} = (\Delta\varphi)_{S,p} > 0. \tag{14.15}$$

By dividing the system in half once again, and introducing a constraint which transfers some entropy from one half to another while the pressure remains constant, we obtain the analogue of Equation (14.9):

$$(\Delta H)_{S\,p} = \left\{\left[\frac{\partial H_1\left(\frac{S}{2}\right)}{\partial S_1}\right]_{p,n/2} - \left[\frac{\partial H_2\left(\frac{S}{2}\right)}{\partial S_2}\right]_{p,n/2}\right\}\Delta S_1 + \frac{1}{2}\left\{\left[\frac{\partial^2 H_1\left(\frac{S}{2}\right)}{\partial S_1^2}\right]_{p,n/2} + \left[\frac{\partial^2 H_2\left(\frac{S}{2}\right)}{\partial S_2^2}\right]_{p,n/2}\right\}(\Delta S_1)^2 + \cdots > 0, \tag{14.16}$$

where in the same manner as with Equations (14.10) and (14.11),

$$\left[\frac{\partial H_1\left(\frac{S}{2}\right)}{\partial S_1}\right]_{p,n/2} = \left[\frac{\partial H_2\left(\frac{S}{2}\right)}{\partial S_2}\right]_{p,n/2} = \left[\frac{\partial H(S)}{\partial S}\right]_{p,n} \tag{14.17}$$

and

$$\left[\frac{\partial^2 H_1\left(\frac{S}{2}\right)}{\partial S_1^2}\right]_{p,n/2} = \left[\frac{\partial^2 H_2\left(\frac{S}{2}\right)}{\partial S_2^2}\right]_{p,n/2} = \left[\frac{\partial^2 H(S)}{\partial S^2}\right]_{p,n} = \left[\frac{\partial T}{\partial S}\right]_{p,n}, \tag{14.18}$$

where Equation (4.80) has been used.

Substitution of Equation (14.17) and (14.18) into Equation (14.16) leads to

$$\left[\frac{\partial S}{\partial T}\right]_{p,n} > 0 \tag{14.19}$$

as a condition of stability. From Equation (6.7) we see that this is equivalent to

$$C_p > 0. \tag{14.20}$$

Thus, the condition of *thermal stability* requires that the heat capacity at constant pressure be positive.

Other stability relations which may be derived by techniques identical with those used in this and the preceding section are:

$$\left[\frac{\partial S}{\partial T}\right]_V > 0 \tag{14.21}$$

and

$$\left[\frac{\partial V}{\partial p}\right]_S < 0. \tag{14.22}$$

4. Compositional Stability

If a single phase consisting of *two* components is considered, it is possible to develop conditions of stability which involve the chemical potential. For this purpose, the Gibbs free energy is employed as a thermodynamic potential. Thus, according to Chapter 8, Section 5, for systems capable of volume work only,

$$(\Delta G)_{T,p} = (\Delta\varphi)_{T,p} > 0. \tag{14.23}$$

Once again, we divide a system (this time consisting of m moles of solvent and n moles of solute) in half. While holding T and p fixed, we introduce a variational constraint such that Δn moles of solute are transferred from one half to the other. In place of Equation (14.9) we get

$$(\Delta G)_{T,p} = \left\{\left[\frac{\partial G_1\left(\frac{n}{2}\right)}{\partial n_1}\right]_{T,p,m/2} - \left[\frac{\partial G_2\left(\frac{n}{2}\right)}{\partial n_2}\right]_{T,p,m/2}\right\}\Delta n_1$$
$$+ \frac{1}{2}\left\{\left[\frac{\partial^2 G_1\left(\frac{n}{2}\right)}{\partial n_1^2}\right]_{T,p,m/2} + \left[\frac{\partial^2 G_2\left(\frac{n}{2}\right)}{\partial n_2^2}\right]_{T,p,m/2}\right\}(\Delta n_1)^2 + \cdots > 0. \tag{14.24}$$

While in place of Equations (14.10) and (14.11),

$$\left[\frac{\partial G_1\left(\frac{n}{2}\right)}{\partial n_1}\right]_{T,p,m/2} = \left[\frac{\partial G_2\left(\frac{n}{2}\right)}{\partial n_2}\right]_{T,p,m/2} = \left[\frac{\partial G(n)}{\partial n}\right]_{T,p,m} \tag{14.25}$$

and

$$\left[\frac{\partial^2 G_1\left(\frac{n}{2}\right)}{\partial n_1^2}\right]_{T,p,m/2} = \left[\frac{\partial^2 G_2\left(\frac{n}{2}\right)}{\partial n_2^2}\right]_{T,p,m/2} = \left[\frac{\partial^2 G(n)}{\partial n^2}\right]_{T,p,m} = \left[\frac{\partial \mu_n}{\partial n}\right]_{T,p,m} \tag{14.26}$$

where Equation (8.43) has been used, and the symbol μ_n has been introduced to denote the chemical potential of the solute. Substitution of Equations (14.25) and (14.26) into Equation (14.24) yields

$$\left[\frac{\partial \mu_n}{\partial n}\right]_{T,p,m} > 0. \tag{14.27}$$

This means that for a stable system, the addition of a component (the mole numbers of other components remaining fixed) results in an increase in the chemical potential of that component, provided that T and p do not change.

If an isothermal plot of chemical potential against mole number n is drawn and an S-shaped region (similar to the $p - V$ isotherm of Figure 14.2) appears, then Equation (14.27) will be violated at some point. When this occurs, we can expect the single, two-component phase to separate into two, so that the phase diagram contains a miscibility gap.

5. Direction of Flow

In Chapter 7, Section 4, it was pointed out that μ, T, and $-p$ were internal potentials in the sense that matter, entropy, and volume would be transferred from regions in which these quantities possessed higher values to regions of lower values, provided that no constraint against the transfer was in force. In this section these statements will be proved.

To begin with μ, if two phases are in contact such that $\mu^{(1)}$ for a given species is higher in phase 1 than $\mu^{(2)}$ for the same species in phase 2, the phase can only be brought into equilibrium when $\mu^{(1)} = \mu^{(2)}$. Since $\mu^{(1)} > \mu^{(2)}$, this means that matter must be transferred in such a manner that $\mu^{(1)}$ decreases while $\mu^{(2)}$ rises to meet it. But according to Equation (14.27), $\mu^{(1)}$ can only decrease if some of the species to which it refers leaves the phase, whereas $\mu^{(2)}$ can only increase if the same species is added to phase 2. Thus, to promote equilibrium, matter leaves the region of higher chemical potential and passes to the region of lower potential.

For the purpose of studying the role of temperature in directing the flow of entropy, we employ the same sort of argument; this time appealing to Equation (14.19). Since temperatures must be equalized at equilibrium, the region of higher temperature must cool while that of lower temperature must rise to meet it. According to Equation (14.19), this can only happen if entropy leaves the high temperature region and enters the one of low temperature.

In order to study the role of $-p$ in directing the flow of volume, the same argument is repeated, using Equation (14.13). We find that volume is transferred from the region of high $-p$ to that of low $-p$.

6. Equilibrium between Drop and Vapor

In Chapter 11 we discussed the equilibrium between a single-component drop and its single-component vapor. We shall now show that this equilibrium is *not* stable. Thus, consider the thermodynamic potential J which, according to Equation (11.18), may be expressed as the sum of $G^{(1)}$ and $R^{(2)}$, where $R^{(2)}$ can be expressed in the form (11.19). Now for stability we require

$$(\Delta J)_{T,p^{(1)}} = \Delta \varphi_{T,p^{(1)}} > 0, \tag{14.28}$$

and it is important to investigate whether or not this is actually so.

Assume that the variational constraint is such that the process to which Equation (14.28) corresponds is the transfer of

$$\Delta n^{(1)} = -\Delta n^{(2)} \tag{14.29}$$

moles of material between the drop and its vapor. During this transfer we have

$$(\Delta J)_{T,p^{(1)}} = (\Delta G^{(1)})_{T,p^{(1)}} + (\Delta R^{(2)})_{T,p^{(1)}}. \tag{14.30}$$

Now we may expand $\Delta G^{(1)}$ and $\Delta R^{(2)}$ in Taylor's series with the results

$$(\Delta G^{(1)})_{T,p^{(1)}} = \left[\frac{\partial G^{(1)}}{\partial n^{(1)}}\right]_{T,p^{(1)}} \Delta n^{(1)} + \frac{1}{2}\left[\frac{\partial^2 G^{(1)}}{\partial n^{(1)2}}\right]_{T,p^{(1)}} (\Delta n^{(1)})^2 + \cdots \tag{14.31}$$

and

$$\Delta R^{(2)}_{T,p^{(1)}} = \left[\frac{\partial R^{(2)}}{\partial n^{(2)}}\right]_{T,p^{(1)}} \Delta n^{(2)} + \frac{1}{2}\left[\frac{\partial^2 R^{(2)}}{\partial n^{(2)^2}}\right]_{T,p^{(1)}} (\Delta n^{(2)})^2 + \cdots \tag{14.32}$$

In these equations the various derivatives refer as usual to the initial state, but in the interest of notational simplicity we have not bothered to symbolize this explicitly.

Now the first derivative on the right of Equation (14.31) is $\mu^{(1)}$ and according to Equation (11.32), in the initial state of equilibrium, this equals the first derivative on the right of Equation (14.32), that is,

$$\left[\frac{\partial G^{(1)}}{\partial n^{(1)}}\right]_{T,p^{(1)}} = \left[\frac{\partial R^{(2)}}{\partial n^{(2)}}\right]_{T,p^{(1)}}. \tag{14.33}$$

Furthermore,

$$\left[\frac{\partial^2 G^{(1)}}{\partial n^{(1)^2}}\right]_{T,p^{(1)}} = \left[\frac{\partial \mu^{(1)}}{\partial n^{(1)}}\right]_{T,p^{(1)}} = 0, \tag{14.34}$$

because the phase consists of only one component, and therefore $\mu^{(1)}$ cannot depend upon $n^{(1)}$ when the intensive variables remain fixed. Upon adding

Equation (14.31) to Equation (14.32) and introducing Equations (14.29), (14.33), and (14.34) into the result we get (when the result is combined with Equation (14.30))

$$(\Delta J)_{T,p^{(1)}} = (\Delta G^{(1)})_{T,p^{(1)}} + (\Delta R^{(2)})_{T,p^{(1)}} = \frac{1}{2}\left[\frac{\partial^2 R^{(2)}}{\partial n^{(2)^2}}\right]_{T,p^{(1)}} (\Delta n^{(2)})^2 + \cdots \tag{14.35}$$

However, according to Equation (11.19),

$$\left[\frac{\partial^2 R^{(2)}}{\partial n^{(2)^2}}\right]_{T,p^{(1)}} = \sigma\left[\frac{\partial^2 \Omega}{\partial n^{(2)^2}}\right]_{T,p^{(1)}} \tag{14.36}$$

The second derivative of the "bulk" term in Equation (11.19) is zero for the same reason that Equation (14.34) is true, while no derivatives of σ appear because we have assumed σ to be independent of drop size. Now the surface area Ω depends upon $(n^{(2)})^{2/3}$ (see Equation (11.47)) and the second derivative of Ω with respect to $n^{(2)}$ is therefore negative. Thus,

$$\left[\frac{\partial^{(2)} R^{(2)}}{\partial n^{(2)^2}}\right]_{T,p^{(1)}} < 0. \tag{14.37}$$

Substitution of Equation (14.37) into Equation (14.35) then gives

$$(\Delta J)_{T,p^{(1)}} < 0 \tag{14.38}$$

in direct contradiction to Equation (14.28). The equilibrium is therefore unstable.

We could have guessed this result from an examination of Equation (11.50), for suppose the radius r of the drop is such that its vapor tension does just equal $p^{(1)}$, that is, suppose that Equation (11.50) is satisfied. Then the transfer of some material from the vapor to the drop will increase r and, according to Equation (11.50), will decrease the vapor tension below $p^{(1)}$. This will cause further condensation and further growth of the drop. On the other hand, suppose material is initially transferred from the drop to the vapor. Then r will have been decreased and, according to Equation (11.50) the vapor tension will exceed $p^{(1)}$. As a result, the drop will continue to evaporate.

Thus, no matter what the direction of the initial displacement, the system will not return automatically to its original state. Therefore, the equilibrium is unstable.

XV

The Third Law

1. Introductory Remarks

We have postponed the discussion of the so-called third law of thermodynamics until this very last chapter. This postponement is a measure of our reluctance to look upon the "third law" as a conventional law of thermodynamics; for it seems that it cannot be used effectively unless some information concerning the atomic and molecular condition of the system is available. In view of this, the third law should be regarded as a useful concept related to thermodynamics while at the same time belonging to a domain which goes beyond its conventional boundaries. This fact should be kept clearly in mind so as not to compromise the power and elegance of the thermodynamic method. On the other hand, because the third law is useful, it cannot and should not be eliminated from consideration. In this chapter, therefore, we shall study the third law and some of its ramifications, taking care to distinguish thermodynamic from nonthermodynamic reasoning.

2. Cooling

Since the third law is inextricably bound to phenomena which take place at low temperatures, it is appropriate to make some remarks concerning how these low temperatures may be achieved. The simplest method for cooling a system is to immerse it in an ambient with a lower temperature than its own. On the other hand, if no such ambient can be found, some means will have to be employed which involves the internal behavior of the system itself. In such instances, it is clear that the most efficient procedure should utilize adiabatic processes so that heat will not flow into the system from its warmer environment.

Consider a simple fluid—a gas—constrained so that it may only perform volume work. One method for lowering its temperature by adiabatic means would be expansion. If S and V are chosen as the independent variables of the system, the differential change in temperature during such an expansion may be expressed as follows:

$$dT = \left[\frac{\partial T}{\partial S}\right]_V dS + \left[\frac{\partial T}{\partial V}\right]_S dV. \tag{15.1}$$

Equations (3.19) and (4.74) permit us to write

$$C_v = \left[\frac{\partial U}{\partial T}\right]_V = T\left[\frac{\partial S}{\partial T}\right]_V. \tag{15.2}$$

Substitution of Equation (15.2) into Equation (15.1) gives

$$dT = \frac{T}{C_v} dS + \left[\frac{\partial T}{\partial V}\right]_S dV. \tag{15.3}$$

The first term on the left of Equation (15.3) is positive, since T is positive and (by the stability arguments of Chapter 14) C_v is also positive. This means that if cooling is involved, the second term in (15.3) must be negative. For a given expansion (a given dV) cooling is therefore maximized when the first term on the left of Equation (15.3) vanishes since it cannot be negative. Since the process is adiabatic, dS is zero when it is conducted reversibly. As a result, maximum cooling is achieved by a *reversible* adiabatic expansion. The same conclusion may be drawn in connection with more general systems which are capable of other kinds of work besides simple volume work; maximum cooling is achieved by a reversible adiabatic process.

In order to cool a gas by adiabatic expansion, it must first be compressed isothermally so that it may subsequently be expanded. As we shall show below in connection with the discussion of adiabatic demagnetization, the greatest reduction in temperature is achieved during the adiabatic process when the greatest reduction of entropy is achieved in the preliminary isothermal process. Reference to Equation (4.92) makes it clear that this reduction in entropy is determined by the thermal expansivity defined by Equation (6.1). Unfortunately, at temperatures below 1°K, it is an empirical fact that thermal expansivities become vanishingly small. As a result, the entropy reduction in the preliminary isothermal process also becomes small; and so cooling by adiabatic expansion is no longer practical.

A more efficient means of cooling involves adiabatic demagnetization. In the presence of a magnetic field, a paramagnetic substance (of constant mass) may be assumed to possess three independent variables of state. These may be chosen as S, V, and $\mathscr{M}$ as in Equation (6.35), or in some other fashion. Suppose we choose the set T, S, and $\mathscr{H}$. Figure 15.1 illustrates the state space predicated on these three variables, and contains a diagram of two different paths corresponding to a change of state leading from state A to state C.

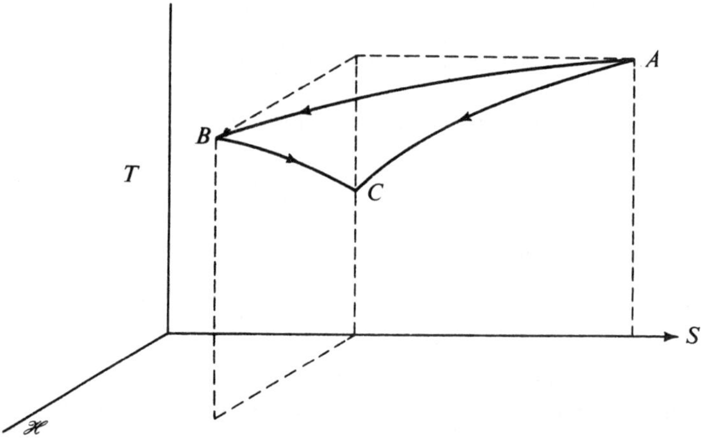

FIGURE 15.1. *Two cooling paths. Path AC is a zero field path, while AB followed by BC consists of isothermal magnetization followed by adiabatic demagnetization.*

We have assumed that C is at a lower temperature than A. Path AC is a zero field path. Thus, it may be imagined that the system initially at temperature T_A, corresponding to state A, is simply immersed in a refrigerant at temperature T_C. Cooling from A to C then takes place in the usual manner. In contrast, the path consisting of AB followed by BC is more detailed. AB is an isothermal path involving an increase of $\mathscr{H}$ and a consequent magnetization of the system. Path BC is isentropic and involves removal of the field. Since BC (having been plotted in state space) is reversible as well as isentropic, Equation (4.1) demands that it also be adiabatic. Thus, BC corresponds to adiabatic demagnetization.

At the moment, we simply assume that a substance can be found which loses entropy (as along AB) when it is magnetized isothermally. To be sure this is possible, it will be necessary to examine the magnetic equation of state of a typical paramagnetic substance.

Paramagnetic salts like $Gd_2(SO_4)_3{\cdot}8H_2O$ obey Curie's law to a high degree of approximation. This law relates the magnetization to the field and has the following form:

$$\mathscr{M} = \frac{\mathscr{H}\gamma}{T}, \tag{15.4}$$

where γ is a positive constant. Substitution of this relation into Equation (6.40) gives

$$\left[\frac{\partial S}{\partial(\mathscr{H}^2)}\right]_{T,p} = -\frac{\gamma}{2T^2}. \tag{15.5}$$

Since γ is positive and relatively large, Equation (15.5) predicts a large decrease of entropy upon isothermal magnetization. Thus, S_B may be considerably less than S_A, and (15.3) predicts cooling. The large value of γ makes magnetic cooling especially useful below 1°K.

The experimental procedure then consists of magnetizing a specimen isothermally (path *AB* of Figure 15.1). When fully magnetized, the specimen is shielded adiabatically and the field is removed reversibly (path *BC* in Figure 15.1). In the first step the entropy of the system is reduced in accordance with Equation (15.5) while in the second step (adiabatic demagnetization) the temperature is reduced.

3. The Nernst Postulate

The third law has been stated in several forms, no two of which are precisely equivalent, and none of which can be utilized without the introduction of some extra-thermodynamic information. The earliest form is the Nernst Postulate, which may be stated as follows:

> At $T = 0$, the entropy change accompanying any process vanishes.

Nernst was led to this statement through a consideration of the possibility of calculating chemical equilibrium constants from purely thermal measurements. From Equation (4.39) it may be seen that the equilibrium constant K depends upon the sum $\sum_i \nu_i \mu_i^\circ$. Now μ_i° represents the Gibbs free energy of one mole of pure species i. Thus, if the molar Gibbs free energies in the pure states of all the species participating in the reaction could be measured by thermal means, one would have in effect measured K also.

From the definitions (4.71) and (3.20) we find

$$\mu_i^\circ = H_i^\circ - TS_i^\circ, \tag{15.6}$$

where H_i° and S_i° are the molar enthalpies and entropies, respectively. In principle, H_i° and S_i° can be determined to within constants through the measurement of heat capacities, followed by application of Formulas (6.33) and (6.34). Unfortunately, the constants of integration remain even if T_0, in these formulas, is set equal to zero. We may write

$$\sum_i \nu_i \mu_i^\circ = \sum_i \nu_i H_i^\circ - T \sum_i \nu_i S_i^\circ. \tag{15.7}$$

The first sum on the right is the negative of the heat of reaction for the pure components, and may be measured by calorimetric means alone. However, the second sum cannot be measured directly by thermal means, but the measurement of heat capacity still represents the best approach. Unfortunately, as we have said, the unknown constants of integration still appear. Thus, if

$$s_i^\circ = \Delta s_i^\circ + \alpha_i, \tag{15.8}$$

where ΔS_i° is that part of the entropy involving integration of the heat capacity and use of latent heat data computed from Equation (6.34), and α_i is the constant of integration (corresponding to $S(T_0,p_0)$ in Equation (6.34)) we have

$$T\sum_i \nu_i s_i^\circ = T\sum_i \nu_i \Delta s_i^\circ + T\sum_i \nu_i\alpha_i. \tag{15.9}$$

The first sum on the right will be known from the heat capacity measurements, but not the second. If $T_0 = 0$, this second sum is clearly the entropy change which accompanies reaction among the pure components at absolute zero.

We have stated repeatedly that constants of integration are of no importance in thermodynamics because they cancel out of all equations which prove useful. Unless some desperate step is taken, Equation (15.7) will prove to be an exception.

In 1906 Nernst took such a step by advancing his postulate. According to this postulate, the last sum in Equation (15.9) must vanish and the above-mentioned dilemma is resolved.

4. Specific Heats and Various Derivatives at Absolute Zero

With $T_0 = 0$ in Equation (6.34), it is clear that the first integral on the right will only converge if $C_p \to 0$ when $T \to 0$. Thus, if the entropy at any finite temperature is to be finite,

$$C_p \to 0 \quad \text{as} \quad T \to 0. \tag{15.10}$$

This argument is entirely independent of the value of $S(0,p_0)$, the entropy at absolute zero. All measurements indicate that C_p does approach zero as the temperature is reduced to zero, and so support the idea that the entropy is finite at finite temperatures.

By a similar argument, it is also possible to suggest that

$$C_v \to 0 \quad \text{as} \quad T \to 0. \tag{15.11}$$

It appears reasonable to adopt the assumption that all heat capacities approach zero as $T \to 0$.

If the Nernst postulate is accepted, one may demonstrate by thermodynamic reasoning that many other derivatives besides the heat capacities vanish as $T \rightarrow 0$. For example, the Nernst postulate demands

$$\left[\frac{\partial S}{\partial V}\right]_T \rightarrow 0 \quad \text{as} \quad T \rightarrow 0, \tag{15.12}$$

$$\left[\frac{\partial S}{\partial p}\right]_T \rightarrow 0 \quad \text{as} \quad T \rightarrow 0, \tag{15.13}$$

since it requires the entropy changes for all isothermal processes to vanish when $T = 0$. From Equations (4.87) and (4.92) we then derive

$$\left[\frac{\partial p}{\partial T}\right]_V \rightarrow 0 \quad \text{as} \quad T \rightarrow 0, \tag{15.14}$$

$$\left[\frac{\partial V}{\partial T}\right]_p \rightarrow 0 \quad \text{as} \quad T \rightarrow 0. \tag{15.15}$$

Other examples, involving more general systems, may be given.

5. The Modified Nernst Postulate

The Nernst postulate was actually stated in stronger fashion by Planck in 1912 who presented it in the following form:

The entropy of any pure substance is zero at absolute zero.

In particular, this requires the Nernst postulate to be true. Unfortunately, subsequent experiments based on Equation (6.34) have invalidated the Planck statement.

Consider a substance which is ordinarily crystalline at low temperatures but which is quenched and cooled in the glassy state. By performing heat capacity measurements on the glass, as well as upon the crystalline solid, one may, through the application of Equation (6.34), determine entropies of both the glass and crystal at some elevated temperature, assuming (in accordance with the Planck statement) that the entropies of both substances are zero at absolute zero. If at the elevated temperature the glass crystallizes, heat is evolved. The entropy change corresponding to this transformation may be measured by dividing the latent heat by the temperature. Thus, the sum of the entropy changes experienced by the glass during heating and crystallizing should equal the entropy change of the solid during heating (since both paths lead to the same thermodynamic state), and by Planck's assumption the entropies of both are initially zero. This predicted experimental result is not observed (the glass has a higher entropy) and so Planck's statement does not seem valid.

Nowadays, we know through the consideration of statistical mechanics that entropy is associated with disorder on an atomic or molecular scale, and so we are not surprised that the entropy of the glass (a disordered structure) is higher than that of the crystal.

As a matter of fact, a restatement of the law has been offered. We shall call this statement the *modified Nernst* postulate. It has the following form:

> The entropy of every pure crystalline solid in its lowest energy state is zero at absolute zero.

The modified Nernst postulate is probably valid (or almost valid); but unfortunately it cannot be employed without using extra-thermodynamic information, for the question of whether or not a solid is ordered, crystalline, and in its lowest energy state, can only be resolved on the basis of atomic and molecular considerations. The fact that the statement goes beyond thermodynamics does not render it valueless. For example, it facilitates determination of the relative entropies of different substances through the application of Equation (6.34). On the other hand, one should clearly distinguish thermodynamic from nonthermodynamic reasoning.

6. The Unattainability of Absolute Zero

The modified Nernst postulate presented in Section 3 of this chapter suffers from the defect that it cannot be used with precision unless some information of a nonmacroscopic sort is available. In an effort to provide a statement of the third law which would be independent of such nonmacroscopic considerations, thermodynamicists have seized upon the empirical fact that further cooling seems to become more difficult as the temperature is lowered. In view of this, the following postulate has been advanced:

> It is impossible by means of any process, no matter how idealized, to reduce the temperature of a system to the absolute zero, in a finite number of steps.

This statement, which is probably valid, is in no way dependent upon nonmacroscopic considerations. However, in its given form, neither is it very valuable for from the experimental viewpoint it is just as useful to be able to approach $T = 0$ with any degree of closeness (when $T = 0$ cannot be precisely attained) as it is to reach $T = 0$. On the other hand, if it proves possible to derive a Nernst-like statement from the unattainability postulate, then the latter is valuable indeed. As we shall demonstrate below, a Nernst-like statement *can* be derived, but only under such special conditions that the resulting theorem amounts to little more than the statement that the entropy change accompanying a process at absolute zero vanishes *when it vanishes* and does not vanish *when it does not vanish.* Thus, not much of real value is gained.

It is important to comprehend this, because some authors seem to insist that the Nernst postulate may be derived from the unattainability postulate. The theorem which can be derived is so heavily qualified that its statement requires an entire paragraph. This paragraph follows.

> Define a "condition" of a system as a selected range of thermodynamic states in which an identifiable variable (for example, volume, magnetic field, chemical constitution, et cetera) is fixed. We then consider a process, $A \rightarrow B$ which carries the system from condition A to condition B. The process will be expansion, change of magnetic state, chemical reaction, et cetera, depending upon which variable is used to define the condition. Suppose that in $A \rightarrow B$ the temperature of the system goes from T' to T'' ($T'' \neq T'$). If the unattainability postulate did not apply, T'' might even be set equal to zero. In this case, suppose the change $A \rightarrow B$ can be conducted in a manner which is both *adiabatic* and *reversible*. We may also consider the reverse process $B \rightarrow A$ in which temperature goes from T'' to T' ($T' \neq T''$), and the special case in which T' is now zero. Suppose that in this instance it is also possible to connect B to A by a path, both *adiabatic* and *reversible*. If all these things are true for both $A \rightarrow B$ and $B \rightarrow A$, and if we then demand that the unattainability postulate hold, it can be proved that for $A \rightarrow B$ or $B \rightarrow A$, conducted at $T = 0$, the accompanying entropy change vanishes, provided that the heat capacities in conditions A and B vanish as $T \rightarrow 0$.

As indicated in Section 4, all heat capacities probably vanish as $T \rightarrow 0$ (this may be considered to be a separate postulate). On the other hand, Equation (4.1) demands that a process which is both adiabatic and reversible is also *isentropic*. Thus, the proof of the above theorem is predicated on the prior knowledge that in *some* process carrying the system from A to B or vice versa (even if it is not the process at $T = 0$), the entropy change is zero. As a result, we find ourselves in the position intimated above of proving that ΔS vanishes in one process if we know that it vanishes in another. As we shall indicate below, it is usually impractical to carry out the first process so that the vanishing or nonvanishing of ΔS cannot be determined by macroscopic measurement, and we are still forced to employ molecular information. However, before carrying this discussion further, we present the proof of the above theorem. This goes as follows.

We may use Equation (6.34) or its analogue for C_v, $C_{\mathscr{H}}$, et cetera, when the variable identifying the "condition" is volume, magnetic field, et cetera, instead of pressure. When the process $A \rightarrow B$ is both adiabatic and reversible and therefore isentropic, we may set the entropy in condition A at T' equal to that in B at T''. According to Equation (6.34) then (assuming that phase transitions do not occur, and that therefore the latent heat terms in (6.34) do not appear), we have

$$S_A(0) + \int_0^{T'} \frac{C_A}{T}\, dT = S_B(0) + \int_0^{T''} \frac{C_B}{T}\, dT. \tag{15.16}$$

In Equation (15.16) the subscripts indicate the "condition" of the system while $S_A(0)$ and $S_B(0)$ are the entropies at $T = 0$. The usual subscripts v, p, $\mathscr{H}$ as in C_v, C_p, and $C_{\mathscr{H}}$ are not shown because we wish to consider the general case. If, for example, C_A and C_B are values of C_v corresponding to different volumes, then $A \rightarrow B$ is an adiabatic expansion in which cooling takes place and $T'' < T'$. However, it is not necessary to introduce this degree of specialization in order to advance the argument.

As long as C_A and C_B pass to zero as $T \rightarrow 0$, the integrals in Equation (15.16) converge. If the system is to be cooled to $T = 0$, we may set $T'' = 0$. Then (15.16) becomes

$$S_B(0) - S_A(0) = \int_0^{T'} \frac{C_A}{T}\, dT. \tag{15.17}$$

Now by the "stability" arguments of Chapter 14, C_A is positive. As a result, Equation (15.17) can be satisfied with a value of $T' > 0$ only if

$$S_B(0) - S_A(0) > 0. \tag{15.18}$$

This means that from an initial state at $T' > 0$, a system has been *cooled* to $T'' = 0$. On the other hand, if the unattainability postulate holds, such a process cannot occur and therefore Equation (15.18) cannot be true. Thus, we are forced to the conclusion, subject to all the qualifying conditions contained in the above statement of the theorem (all of which have been used in the proof) that

$$S_B(0) - S_A(0) \leq 0, \tag{15.19}$$

a condition which excludes Equation (15.18).

By performing the argument precisely in reverse, using the process $B \rightarrow A$, we obtain

$$S_A(0) - S_B(0) \leq 0. \tag{15.20}$$

Conditions (15.19) and (15.20) are only consistent if

$$S_A(0) = S_B(0). \tag{15.21}$$

Thus, for the process $A \rightleftarrows B$, subject to all the above-mentioned qualifying conditions, the entropy change vanishes at $T = 0$. This completes the proof.

An inverse proof may also be given which shows that if Equation (15.21) is true, then none of the above-mentioned, highly qualified processes can be used to cool the system to absolute zero. For example, if the left-hand side of Equation (15.17) is set equal to zero, then no $T' > 0$ can be found (since C_A is positive) which satisfies the resulting equation, and therefore no initial temperature T' exists from which the system can be cooled by one of the qualified processes to $T'' = 0$.

Processes $A \rightarrow B$ do exist but they cannot be conducted both adiabatically and reversibly. For example, if A is a glass at a temperature T' and B is the crystalline form of the species comprising the glass, but at $T'' > T'$, then it is an empirical fact that the transformation will not be isentropic (and therefore cannot be performed in a manner both adiabatic and reversible), but involves a decrease in entropy. Thus, Equation (15.16) does not hold for this change of "condition." It is fortunate that this is so or else the modified Nernst postulate which seems to be valid, would have to be replaced by the original Nernst postulate, or else the unattainability statement would have to be relinquished.

Some authors make it a point to ascribe the failure to achieve the result (15.21) in the example of the glass to the impossibility of carrying out the glass-to-crystal transformation *reversibly*, rather than focussing upon the incompatibility of "reversible" with "adiabatic," a point which has been emphasized here. This erroneous outlook stems from the view that the glass is not in ordinary equilibrium but "metastable" equilibrium, and therefore cannot be transformed reversibly to the crystalline condition. The discussion of Chapter 1, Section 9 makes it clear that the concept of metastability has no "formal" place in classical thermodynamic argument, and that by employing suitable constraints and a sufficient number of variables the process can indeed be performed reversibly. This in no way damages the argument of the present chapter, because it is the combination of both "reversible" and "adiabatic" which holds the key, and not just "reversible" itself.

In the case of the glass, we could not easily perform macroscopic measurements which would allow us in advance to determine whether $A \rightarrow B$ (going from $T' > 0$ to $T'' = 0$) is isentropic. For one thing, in the face of the unattainability postulate we could not even carry out the processes because $T'' = 0$ could not be reached. Our only recourse then would involve an appeal to information concerning the degree of molecular or atomic ordering. But then we would be in the same position as with the modified Nernst statement. Thus, the unattainability postulate, though itself wholly macroscopic, is of no help.

The modified Nernst postulate is of real value because it provides us with information on the relations between entropy integration constants, and therefore permits the determination of relative entropies. However, it cannot be used effectively without information of a nonmacroscopic sort. We must therefore accept the third law as a useful concept, but one which is not sustained by thermodynamics alone. It should not therefore be regarded as a law of thermodynamics.

Index

Absolute zero, 8
  unattainability, 210–14
Adiabatic, 33
  demagnetization, 205–206
  wall, 33
  work, 37
Adiabatically inaccessible states, 38–41

Barometric formula, 184
Boiling point elevation, 140–42

Caratheodory principle, 71–77
Caratheodory theorem, 22–26
Carnot cycle, 63
Centrifugal field, 178–86
Chemical potential, 113, 127–29
  of drop, 163–65
  in elastic system, 190–91
  in fields, 180–82
  in ideal gas, 136–37
  and variable of state, 136–37
Clapeyron-Clausius equation, 137–39
Clausius principle, 70
Closed system, 43
Compression, 34
Compressibility, isothermal, 99
Concentrated cell, 148
Conjugate variables, 82
Conversion of heat to work, 69–70
Constraints, variables, and work, 11–16
Cooling, 204–207

Dalton's law of partial pressures, 90
Deductive frameworks, 18–19
Degree of constraint, 54–56
Degrees of freedom, 132–33
Diathermic wall, 34
Dilute solution, 144–46
Direction of change, 127
Direction of flow, 201
Distribution in centrifuge, 185–86

Efficiency, 64
  engineering, 77
  maximum, 65–68
  thermodynamic, 64–65
Elastic systems, 187–93
  vapor pressure, 191–93
Electrochemical cells, 146–49
Enthalpy, 43–44
  of mixing, 93
Entropy, 50–60
  from heat capacity, 106
  of ideal gas, 89–90
Equation of state, 9, 36, 82–83
Equilibrium, 3
  constant, 143
  feature of, 112–13
  metastable, 16–18, 213
  stable, 194
  unstable, 18, 203
Euler's theorem, 28
Exact differential, 20–21
Expansion, 34
Extensive quantities, 46–47

Extensive variable, 29
Extremal, 29
  condition for entropy, 108
  conditions for thermodynamic potentials, 122–25
  constrained, 29–31

First law of thermodynamics, 41–43
Free expansion, 101–103
Freezing point depression, 142
Function of state, 43
  derived, 43
  primary, 43

Generalized feature of equilibrium, 125–27, 150–51
Gibbs adsorption isotherm, 177
Gibbs-Duhem relation, 134
  for surface, 175–76
Gibbs free energy, 78–79
Glass, 209–10, 212–13
Gravitational field, 178–86

Heat, 42
Heat capacity, 43–44
  of ideal gas, 88
  at constant pressure, 44
  at constant volume, 43
  and measurement of enthalpy and entropy, 106
Heat content, 43–44
Heat of reaction, 49
Helmholtz free energy, 78
Homogeneous functions, 28–29

Ideal gas, 83–93
Ideal solution, 93–98
  partial molar free energy of, 95–97
Independently variable components, 132–33
Induced process, 53
Initial constraints, 118
Integrating denominator, 21
Intensive variables, 29, 46–47
Internal energy, 41–43
  of ideal gas, 88
Internal equilibrium, 109–12
  conditions of, 109–12, 129–32
  elementary method for, 116–18
  potentials for, 113–14, 201

Inverse process, 59
Inversion temperature, 104
Irreversible process, 53
Irreversible thermodynamics, 2
Isobaric change, 78
Isolated system, 43
Isothermal change, 9
Isotherms, 36

Joule-Thomson coefficient, 103–105
Junction potential, 148

Lagrange method of undetermined multipliers, 29–31
Laplace relation, 157–60
Legendre transformation, 79–81

Macroscopic state space, 5–6
Macroscopic systems, 1
Mass action, law of, 142–43
Maxwell relations, 79–81
Mechanical system, 52
Metastability, 213
Microscopic system, 1
Mole fraction, 92
Mole numbers, 45

Nernst postulate, 207–208
  modified, 209–10
Null complex, 23

Organization of system, 56
Osmotic pressure, 154
Osmotic system, 151–55

Partial molar quantities, 47–48
Partial pressure, 90
Pfaff differential equation, 22
  solution curve of, 22
Pfaff differential expression, 21
Phase, 45
  open, 110
  rule, 132–33
  surface, 169
Phenomenology, 2

Quasistatic process, 9–11

Raoult's law, 139–40
Reciprocity relations, 21

Removal of constraint, 12
Reversible environment, 66, 68–69, 122
Reversible process, 9–11

Second law of thermodynamics, 50
Stability, 194–203
  compositional, 200–201
  mechanical, 195–98
  of drop, 202–203
  thermal, 199–200
Steady state, 3
Stirring, 34
Strain, 187–88
Stress, 187–88
Superficial density, 169
Surface
  boundary between liquid and vapor, 45
  concentration, 169
  constraints, 170–72
  excess quantities, 169–70
  Gibbs dividing, 168
  layers, 156–57
  phase, 169
  solubility, effect on, 167
  structure, 167–77
  tension, 156, 157, 172
  thermodynamic potential, 160–63, 172–74
  vapor pressure, effect on, 165–67

Temperature, 33–37
  Centigrade, 84
  empirical, 36
  Kelvin, 65, 85
  relation between Kelvin and thermodynamic, 86–87
  thermodynamic, 37, 62–65
Thermal diffusion, 118
Thermal expansivity, 99
Thermodynamic activity, 143–45
  measurement, 149
Thermodynamic change of state, 6
Thermodynamic potential, 115–27
  elastic system, 190
  open system, 134–35
  surface system, 160–63, 172–74
Thermodynamic process, 6
Thermodynamic relations between magnetic quantities, 106–107
Thermodynamic state, 4–5
  reproducibility, 3–4
Thermodynamic system, 2
Thermometer, 36
Thermo-osmosis, 118
Thermostatics, 2
Third law of thermodynamics, 204–14

Van der Waals fluid, 198–99
Variables of state, 4–5
  geometric, 76
  extensive, 29, 46–47
  intensive, 29, 46–47
  transformation, 26
  variance, 132–33
Variational constraint, 118
Virtual variation, 60–62
Volume of mixing, 97–98

Work
  content, 78
  elastic, 7
  electrical, 7
  magnetic, 7–8
  mechanical, 6–7
  surface, 7
  volume, 6–7

Zeroth law of thermodynamics, 32

A CATALOG OF SELECTED

DOVER BOOKS

IN SCIENCE AND MATHEMATICS

Astronomy

BURNHAM'S CELESTIAL HANDBOOK, Robert Burnham, Jr. Thorough guide to the stars beyond our solar system. Exhaustive treatment. Alphabetical by constellation: Andromeda to Cetus in Vol. 1; Chamaeleon to Orion in Vol. 2; and Pavo to Vulpecula in Vol. 3. Hundreds of illustrations. Index in Vol. 3. 2,000pp. 6⅛ x 9¼.
Vol. I: 23567-X
Vol. II: 23568-8
Vol. III: 23673-0

EXPLORING THE MOON THROUGH BINOCULARS AND SMALL TELESCOPES, Ernest H. Cherrington, Jr. Informative, profusely illustrated guide to locating and identifying craters, rills, seas, mountains, other lunar features. Newly revised and updated with special section of new photos. Over 100 photos and diagrams. 240pp. 8¼ x 11. 24491-1

THE EXTRATERRESTRIAL LIFE DEBATE, 1750–1900, Michael J. Crowe. First detailed, scholarly study in English of the many ideas that developed from 1750 to 1900 regarding the existence of intelligent extraterrestrial life. Examines ideas of Kant, Herschel, Voltaire, Percival Lowell, many other scientists and thinkers. 16 illustrations. 704pp. 5⅜ x 8½. 40675-X

THEORIES OF THE WORLD FROM ANTIQUITY TO THE COPERNICAN REVOLUTION, Michael J. Crowe. Newly revised edition of an accessible, enlightening book recreates the change from an earth-centered to a sun-centered conception of the solar system. 242pp. 5⅜ x 8½. 41444-2

A HISTORY OF ASTRONOMY, A. Pannekoek. Well-balanced, carefully reasoned study covers such topics as Ptolemaic theory, work of Copernicus, Kepler, Newton, Eddington's work on stars, much more. Illustrated. References. 521pp. 5⅜ x 8½.
65994-1

A COMPLETE MANUAL OF AMATEUR ASTRONOMY: Tools and Techniques for Astronomical Observations, P. Clay Sherrod with Thomas L. Koed. Concise, highly readable book discusses: selecting, setting up and maintaining a telescope; amateur studies of the sun; lunar topography and occultations; observations of Mars, Jupiter, Saturn, the minor planets and the stars; an introduction to photoelectric photometry; more. 1981 ed. 124 figures. 26 halftones. 37 tables. 335pp. 6½ x 9¼.
42820-6

AMATEUR ASTRONOMER'S HANDBOOK, J. B. Sidgwick. Timeless, comprehensive coverage of telescopes, mirrors, lenses, mountings, telescope drives, micrometers, spectroscopes, more. 189 illustrations. 576pp. 5⅜ x 8¼. (Available in U.S. only.)
24034-7

STARS AND RELATIVITY, Ya. B. Zel'dovich and I. D. Novikov. Vol. 1 of *Relativistic Astrophysics* by famed Russian scientists. General relativity, properties of matter under astrophysical conditions, stars, and stellar systems. Deep physical insights, clear presentation. 1971 edition. References. 544pp. 5⅜ x 8¼. 69424-0

Chemistry

THE SCEPTICAL CHYMIST: The Classic 1661 Text, Robert Boyle. Boyle defines the term "element," asserting that all natural phenomena can be explained by the motion and organization of primary particles. 1911 ed. viii+232pp. 5⅜ x 8½. 42825-7

RADIOACTIVE SUBSTANCES, Marie Curie. Here is the celebrated scientist's doctoral thesis, the prelude to her receipt of the 1903 Nobel Prize. Curie discusses establishing atomic character of radioactivity found in compounds of uranium and thorium; extraction from pitchblende of polonium and radium; isolation of pure radium chloride; determination of atomic weight of radium; plus electric, photographic, luminous, heat, color effects of radioactivity. ii+94pp. 5⅜ x 8½. 42550-9

CHEMICAL MAGIC, Leonard A. Ford. Second Edition, Revised by E. Winston Grundmeier. Over 100 unusual stunts demonstrating cold fire, dust explosions, much more. Text explains scientific principles and stresses safety precautions. 128pp. 5⅜ x 8½. 67628-5

THE DEVELOPMENT OF MODERN CHEMISTRY, Aaron J. Ihde. Authoritative history of chemistry from ancient Greek theory to 20th-century innovation. Covers major chemists and their discoveries. 209 illustrations. 14 tables. Bibliographies. Indices. Appendices. 851pp. 5⅜ x 8½. 64235-6

CATALYSIS IN CHEMISTRY AND ENZYMOLOGY, William P. Jencks. Exceptionally clear coverage of mechanisms for catalysis, forces in aqueous solution, carbonyl- and acyl-group reactions, practical kinetics, more. 864pp. 5⅜ x 8½. 65460-5

ELEMENTS OF CHEMISTRY, Antoine Lavoisier. Monumental classic by founder of modern chemistry in remarkable reprint of rare 1790 Kerr translation. A must for every student of chemistry or the history of science. 539pp. 5⅜ x 8½. 64624-6

THE HISTORICAL BACKGROUND OF CHEMISTRY, Henry M. Leicester. Evolution of ideas, not individual biography. Concentrates on formulation of a coherent set of chemical laws. 260pp. 5⅜ x 8½. 61053-5

A SHORT HISTORY OF CHEMISTRY, J. R. Partington. Classic exposition explores origins of chemistry, alchemy, early medical chemistry, nature of atmosphere, theory of valency, laws and structure of atomic theory, much more. 428pp. 5⅜ x 8½. (Available in U.S. only.) 65977-1

GENERAL CHEMISTRY, Linus Pauling. Revised 3rd edition of classic first-year text by Nobel laureate. Atomic and molecular structure, quantum mechanics, statistical mechanics, thermodynamics correlated with descriptive chemistry. Problems. 992pp. 5⅜ x 8½. 65622-5

FROM ALCHEMY TO CHEMISTRY, John Read. Broad, humanistic treatment focuses on great figures of chemistry and ideas that revolutionized the science. 50 illustrations. 240pp. 5⅜ x 8½. 28690-8

Engineering

DE RE METALLICA, Georgius Agricola. The famous Hoover translation of greatest treatise on technological chemistry, engineering, geology, mining of early modern times (1556). All 289 original woodcuts. 638pp. 6¾ x 11. 60006-8

FUNDAMENTALS OF ASTRODYNAMICS, Roger Bate et al. Modern approach developed by U.S. Air Force Academy. Designed as a first course. Problems, exercises. Numerous illustrations. 455pp. 5⅜ x 8½. 60061-0

DYNAMICS OF FLUIDS IN POROUS MEDIA, Jacob Bear. For advanced students of ground water hydrology, soil mechanics and physics, drainage and irrigation engineering, and more. 335 illustrations. Exercises, with answers. 784pp. 6⅛ x 9¼. 65675-6

THEORY OF VISCOELASTICITY (Second Edition), Richard M. Christensen. Complete, consistent description of the linear theory of the viscoelastic behavior of materials. Problem-solving techniques discussed. 1982 edition. 29 figures. xiv+364pp. 6⅛ x 9¼. 42880-X

MECHANICS, J. P. Den Hartog. A classic introductory text or refresher. Hundreds of applications and design problems illuminate fundamentals of trusses, loaded beams and cables, etc. 334 answered problems. 462pp. 5⅜ x 8½. 60754-2

MECHANICAL VIBRATIONS, J. P. Den Hartog. Classic textbook offers lucid explanations and illustrative models, applying theories of vibrations to a variety of practical industrial engineering problems. Numerous figures. 233 problems, solutions. Appendix. Index. Preface. 436pp. 5⅜ x 8½. 64785-4

STRENGTH OF MATERIALS, J. P. Den Hartog. Full, clear treatment of basic material (tension, torsion, bending, etc.) plus advanced material on engineering methods, applications. 350 answered problems. 323pp. 5⅜ x 8½. 60755-0

A HISTORY OF MECHANICS, René Dugas. Monumental study of mechanical principles from antiquity to quantum mechanics. Contributions of ancient Greeks, Galileo, Leonardo, Kepler, Lagrange, many others. 671pp. 5⅜ x 8½. 65632-2

STABILITY THEORY AND ITS APPLICATIONS TO STRUCTURAL MECHANICS, Clive L. Dym. Self-contained text focuses on Koiter postbuckling analyses, with mathematical notions of stability of motion. Basing minimum energy principles for static stability upon dynamic concepts of stability of motion, it develops asymptotic buckling and postbuckling analyses from potential energy considerations, with applications to columns, plates, and arches. 1974 ed. 208pp. 5⅜ x 8½. 42541-X

METAL FATIGUE, N. E. Frost, K. J. Marsh, and L. P. Pook. Definitive, clearly written, and well-illustrated volume addresses all aspects of the subject, from the historical development of understanding metal fatigue to vital concepts of the cyclic stress that causes a crack to grow. Includes 7 appendixes. 544pp. 5⅜ x 8½. 40927-9

Math–Decision Theory, Statistics, Probability

ELEMENTARY DECISION THEORY, Herman Chernoff and Lincoln E. Moses. Clear introduction to statistics and statistical theory covers data processing, probability and random variables, testing hypotheses, much more. Exercises. 364pp. 5⅜ x 8½. 65218-1

STATISTICS MANUAL, Edwin L. Crow et al. Comprehensive, practical collection of classical and modern methods prepared by U.S. Naval Ordnance Test Station. Stress on use. Basics of statistics assumed. 288pp. 5⅜ x 8½. 60599-X

SOME THEORY OF SAMPLING, William Edwards Deming. Analysis of the problems, theory, and design of sampling techniques for social scientists, industrial managers, and others who find statistics important at work. 61 tables. 90 figures. xvii +602pp. 5⅜ x 8½. 64684-X

LINEAR PROGRAMMING AND ECONOMIC ANALYSIS, Robert Dorfman, Paul A. Samuelson and Robert M. Solow. First comprehensive treatment of linear programming in standard economic analysis. Game theory, modern welfare economics, Leontief input-output, more. 525pp. 5⅜ x 8½. 65491-5

PROBABILITY: An Introduction, Samuel Goldberg. Excellent basic text covers set theory, probability theory for finite sample spaces, binomial theorem, much more. 360 problems. Bibliographies. 322pp. 5⅜ x 8½. 65252-1

GAMES AND DECISIONS: Introduction and Critical Survey, R. Duncan Luce and Howard Raiffa. Superb nontechnical introduction to game theory, primarily applied to social sciences. Utility theory, zero-sum games, n-person games, decision-making, much more. Bibliography. 509pp. 5⅜ x 8½. 65943-7

INTRODUCTION TO THE THEORY OF GAMES, J. C. C. McKinsey. This comprehensive overview of the mathematical theory of games illustrates applications to situations involving conflicts of interest, including economic, social, political, and military contexts. Appropriate for advanced undergraduate and graduate courses; advanced calculus a prerequisite. 1952 ed. x+372pp. 5⅜ x 8½. 42811-7

FIFTY CHALLENGING PROBLEMS IN PROBABILITY WITH SOLUTIONS, Frederick Mosteller. Remarkable puzzlers, graded in difficulty, illustrate elementary and advanced aspects of probability. Detailed solutions. 88pp. 5⅜ x 8½. 65355-2

PROBABILITY THEORY: A Concise Course, Y. A. Rozanov. Highly readable, self-contained introduction covers combination of events, dependent events, Bernoulli trials, etc. 148pp. 5⅜ x 8¼. 63544-9

STATISTICAL METHOD FROM THE VIEWPOINT OF QUALITY CONTROL, Walter A. Shewhart. Important text explains regulation of variables, uses of statistical control to achieve quality control in industry, agriculture, other areas. 192pp. 5⅜ x 8½. 65232-7

Math–Geometry and Topology

ELEMENTARY CONCEPTS OF TOPOLOGY, Paul Alexandroff. Elegant, intuitive approach to topology from set-theoretic topology to Betti groups; how concepts of topology are useful in math and physics. 25 figures. 57pp. 5⅜ x 8½. 60747-X

COMBINATORIAL TOPOLOGY, P. S. Alexandrov. Clearly written, well-organized, three-part text begins by dealing with certain classic problems without using the formal techniques of homology theory and advances to the central concept, the Betti groups. Numerous detailed examples. 654pp. 5⅜ x 8½. 40179-0

EXPERIMENTS IN TOPOLOGY, Stephen Barr. Classic, lively explanation of one of the byways of mathematics. Klein bottles, Moebius strips, projective planes, map coloring, problem of the Koenigsberg bridges, much more, described with clarity and wit. 43 figures. 210pp. 5⅜ x 8½. 25933-1

CONFORMAL MAPPING ON RIEMANN SURFACES, Harvey Cohn. Lucid, insightful book presents ideal coverage of subject. 334 exercises make book perfect for self-study. 55 figures. 352pp. 5⅝ x 8¼. 64025-6

THE GEOMETRY OF RENÉ DESCARTES, René Descartes. The great work founded analytical geometry. Original French text, Descartes's own diagrams, together with definitive Smith-Latham translation. 244pp. 5⅜ x 8½. 60068-8

PRACTICAL CONIC SECTIONS: The Geometric Properties of Ellipses, Parabolas and Hyperbolas, J. W. Downs. This text shows how to create ellipses, parabolas, and hyperbolas. It also presents historical background on their ancient origins and describes the reflective properties and roles of curves in design applications. 1993 ed. 98 figures. xii+100pp. 6½ x 9¼. 42876-1

THE THIRTEEN BOOKS OF EUCLID'S ELEMENTS, translated with introduction and commentary by Thomas L. Heath. Definitive edition. Textual and linguistic notes, mathematical analysis. 2,500 years of critical commentary. Unabridged. 1,414pp. 5⅜ x 8½. Three-vol. set. Vol. I: 60088-2 Vol. II: 60089-0 Vol. III: 60090-4

GEOMETRY OF COMPLEX NUMBERS, Hans Schwerdtfeger. Illuminating, widely praised book on analytic geometry of circles, the Moebius transformation, and two-dimensional non-Euclidean geometries. 200pp. 5⅝ x 8¼. 63830-8

DIFFERENTIAL GEOMETRY, Heinrich W. Guggenheimer. Local differential geometry as an application of advanced calculus and linear algebra. Curvature, transformation groups, surfaces, more. Exercises. 62 figures. 378pp. 5⅜ x 8½. 63433-7

CURVATURE AND HOMOLOGY: Enlarged Edition, Samuel I. Goldberg. Revised edition examines topology of differentiable manifolds; curvature, homology of Riemannian manifolds; compact Lie groups; complex manifolds; curvature, homology of Kaehler manifolds. New Preface. Four new appendixes. 416pp. 5⅜ x 8½. 40207-X

History of Math

THE WORKS OF ARCHIMEDES, Archimedes (T. L. Heath, ed.). Topics include the famous problems of the ratio of the areas of a cylinder and an inscribed sphere; the measurement of a circle; the properties of conoids, spheroids, and spirals; and the quadrature of the parabola. Informative introduction. clxxxvi+326pp; supplement, 52pp. 5⅜ x 8½. 42084-1

A SHORT ACCOUNT OF THE HISTORY OF MATHEMATICS, W. W. Rouse Ball. One of clearest, most authoritative surveys from the Egyptians and Phoenicians through 19th-century figures such as Grassman, Galois, Riemann. Fourth edition. 522pp. 5⅜ x 8½. 20630-0

THE HISTORY OF THE CALCULUS AND ITS CONCEPTUAL DEVELOPMENT, Carl B. Boyer. Origins in antiquity, medieval contributions, work of Newton, Leibniz, rigorous formulation. Treatment is verbal. 346pp. 5⅜ x 8½. 60509-4

THE HISTORICAL ROOTS OF ELEMENTARY MATHEMATICS, Lucas N. H. Bunt, Phillip S. Jones, and Jack D. Bedient. Fundamental underpinnings of modern arithmetic, algebra, geometry, and number systems derived from ancient civilizations. 320pp. 5⅜ x 8½. 25563-8

A HISTORY OF MATHEMATICAL NOTATIONS, Florian Cajori. This classic study notes the first appearance of a mathematical symbol and its origin, the competition it encountered, its spread among writers in different countries, its rise to popularity, its eventual decline or ultimate survival. Original 1929 two-volume edition presented here in one volume. xxviii+820pp. 5⅜ x 8½. 67766-4

GAMES, GODS & GAMBLING: A History of Probability and Statistical Ideas, F. N. David. Episodes from the lives of Galileo, Fermat, Pascal, and others illustrate this fascinating account of the roots of mathematics. Features thought-provoking references to classics, archaeology, biography, poetry. 1962 edition. 304pp. 5⅜ x 8½. (Available in U.S. only.) 40023-9

OF MEN AND NUMBERS: The Story of the Great Mathematicians, Jane Muir. Fascinating accounts of the lives and accomplishments of history's greatest mathematical minds–Pythagoras, Descartes, Euler, Pascal, Cantor, many more. Anecdotal, illuminating. 30 diagrams. Bibliography. 256pp. 5⅜ x 8½. 28973-7

HISTORY OF MATHEMATICS, David E. Smith. Nontechnical survey from ancient Greece and Orient to late 19th century; evolution of arithmetic, geometry, trigonometry, calculating devices, algebra, the calculus. 362 illustrations. 1,355pp. 5⅜ x 8½. Two-vol. set. Vol. I: 20429-4 Vol. II: 20430-8

A CONCISE HISTORY OF MATHEMATICS, Dirk J. Struik. The best brief history of mathematics. Stresses origins and covers every major figure from ancient Near East to 19th century. 41 illustrations. 195pp. 5⅜ x 8½. 60255-9

Physics

OPTICAL RESONANCE AND TWO-LEVEL ATOMS, L. Allen and J. H. Eberly. Clear, comprehensive introduction to basic principles behind all quantum optical resonance phenomena. 53 illustrations. Preface. Index. 256pp. 5⅜ x 8½. 65533-4

QUANTUM THEORY, David Bohm. This advanced undergraduate-level text presents the quantum theory in terms of qualitative and imaginative concepts, followed by specific applications worked out in mathematical detail. Preface. Index. 655pp. 5⅜ x 8½. 65969-0

ATOMIC PHYSICS: 8th edition, Max Born. Nobel laureate's lucid treatment of kinetic theory of gases, elementary particles, nuclear atom, wave-corpuscles, atomic structure and spectral lines, much more. Over 40 appendices, bibliography. 495pp. 5⅜ x 8½. 65984-4

A SOPHISTICATE'S PRIMER OF RELATIVITY, P. W. Bridgman. Geared toward readers already acquainted with special relativity, this book transcends the view of theory as a working tool to answer natural questions: What is a frame of reference? What is a "law of nature"? What is the role of the "observer"? Extensive treatment, written in terms accessible to those without a scientific background. 1983 ed. xlviii+172pp. 5⅜ x 8½. 42549-5

AN INTRODUCTION TO HAMILTONIAN OPTICS, H. A. Buchdahl. Detailed account of the Hamiltonian treatment of aberration theory in geometrical optics. Many classes of optical systems defined in terms of the symmetries they possess. Problems with detailed solutions. 1970 edition. xv+360pp. 5⅜ x 8½. 67597-1

PRIMER OF QUANTUM MECHANICS, Marvin Chester. Introductory text examines the classical quantum bead on a track: its state and representations; operator eigenvalues; harmonic oscillator and bound bead in a symmetric force field; and bead in a spherical shell. Other topics include spin, matrices, and the structure of quantum mechanics; the simplest atom; indistinguishable particles; and stationary-state perturbation theory. 1992 ed. xiv+314pp. 6⅛ x 9¼. 42878-8

LECTURES ON QUANTUM MECHANICS, Paul A. M. Dirac. Four concise, brilliant lectures on mathematical methods in quantum mechanics from Nobel Prize–winning quantum pioneer build on idea of visualizing quantum theory through the use of classical mechanics. 96pp. 5⅜ x 8½. 41713-1

THIRTY YEARS THAT SHOOK PHYSICS: The Story of Quantum Theory, George Gamow. Lucid, accessible introduction to influential theory of energy and matter. Careful explanations of Dirac's anti-particles, Bohr's model of the atom, much more. 12 plates. Numerous drawings. 240pp. 5⅜ x 8½. 24895-X

ELECTRONIC STRUCTURE AND THE PROPERTIES OF SOLIDS: The Physics of the Chemical Bond, Walter A. Harrison. Innovative text offers basic understanding of the electronic structure of covalent and ionic solids, simple metals, transition metals and their compounds. Problems. 1980 edition. 582pp. 6⅛ x 9¼. 66021-4